Statistical Computing in Pascal

D Cooke
A H Craven
G M Clarke
Mathematics Division, University of Sussex

Edward Arnold

First published 1985 by
Edward Arnold (Publishers) Ltd
41 Bedford Square, London WC1B 3DQ

Edward Arnold, 300 North Charles Street,
Baltimore, Maryland 21201, USA

Edward Arnold (Australia) Pty Ltd,
80 Waverley Road, Caulfield East,
Victoria 3145, Australia

British Library Cataloguing in Publication Data

Cooke, D.
   Statistical computing in Pascal.
   1. Mathematical statistics—Data processing
   2. PASCAL (Computer program language)
   I. Title   II. Craven, A. H.   III. Clarke, G. M.
   519.5'028'5424   QA276.A

ISBN 0-7131-3545-X

Printed in Great Britain by
Thomson Litho Ltd, East Kilbride, Scotland

Preface

Our aim in writing this book has been to show how a computer can best be used to help analyse statistical data, especially for readers who are not highly expert in programming. We thus emphasise simple, fundamental procedures of statistics, and for descriptions of these we frequently refer to the text *A Basic Course in Statistics*, by G M Clarke and D Cooke (Edward Arnold, Second Edition 1983): references are given in the form '*ABC*, Section 0.0'.

The programs in this book are written in Pascal, a language that is increasing in popularity and may now be the first language learnt by many students. (Our earlier book, *Basic Statistical Computing*, Edward Arnold, 1982, contained programs written in BASIC, another widely-used language.) We do not attempt to teach Pascal, and the reader will need an elementary knowledge of it, say at the level of what is often called 'simple Pascal'. With this background, the reader should easily be able to learn additional features of the language as they occur. We have tried to conform to the original specification of Pascal, in the hope that our algorithms can be implemented on any machine on which Pascal can be used.

The student of statistics will find a repertoire of algorithms that will help him not only with the calculations needed in data analysis but also with exercises that give him a deeper insight into theoretical statistics. A wide range of programming techniques can be applied to statistical problems; the student of computing will therefore find that the statistics context provides an interesting background in which to develop programming skills. Pascal was designed as a language that gives programs with a strong structure, and this feature is extremely useful in writing the modular, flexible programs that are appropriate for analysing statistical data.

We have included chapters on regression and analysis of variance, because these are methods needed widely by users of statistics. This text will thus serve research workers in many fields as a reference collection of computing algorithms for basic statistical methods.

In Appendix B, there is a large set of data, parts of which can be used in various exercises throughout the text: suitable exercises are marked†. Additional suggestions for use of the data are given on page 164.

We are very grateful to Mr W A Craven, whose expertise in Pascal has been of invaluable help in developing algorithms. It is, as always, a pleasure to thank Mrs Jill Foster for her assistance in preparing typescript.

D Cooke
A H Craven
G M Clarke

Sussex
August 1984

Contents

List of Algorithms and Programs

Type Declarations

Values of constants are given as illustrations; appropriate values should be chosen for particular applications.

Declarations which are used only in the analysis of variance are given on page 132.

Data

```
const    MaxSampleSize = 100;
         MaxVariate = 10;
type     Units = 1. .MaxSampleSize;
         Variates = 1. .MaxVariate;
         DataVector = array[Units] of real;
         DataList = array[Units] of integer;
         DataMatrix = array[Variates] of DataVector;
         DataTable = array[Variates] of DataList;
```

Frequency tables

```
const    MaxClass = 10;
         MaxCol = 3;
         MaxRow = 3;
type     Classes = 0. .MaxClass;
         Cols = 1. .MaxCol;
         Rows = 1. .MaxRow;
         FreqVector = array[Classes] of real;
         FreqList = array[Classes] of integer;
         FreqMatrix = array[Rows,Cols] of real;
         FreqTable = array[Rows,Cols] of integer;
```

Miscellaneous

```
const    MaxArraySize = 28;
type     Vector = array[Variates] of real;
         List = array[Variates] of integer;
         MatrixArray = array[1. .MaxArraySize] of real;
         QuartileArray = array[0. .4] of real;
```

1 Introduction

1.1 The development of calculating devices in statistics

The rapid growth of statistics is a feature of the twentieth century. It has benefited from the development of more powerful calculating machines and computers. By 1900, precision engineering was far enough advanced for efficient calculators to be produced at attractive prices—the eminent statistician Karl Pearson wrote in a letter of 1894: 'I want to purchase a Brunsviga calculating machine before anything else, and am making inquiries about it. I think it would make moment-calculating easy.' Another development of the time was due to Hollerith, in the United States, who wanted to automate the tabulation of census data. He represented data by holes punched in cards, and devised electrical machines to sort and count the punched cards: these were used in the 1890 Census in the USA. Hollerith's company for manufacturing these machines has become part of International Business Machines (IBM), well-known today as computer manufacturers.

In the early 1900s, routine statistical analysis mostly involved tabulating data and calculating averages and index numbers. Simple adding and listing machines then used could add and subtract, print out totals, sub-totals, and individual items if required; Hollerith's punched-card machines did this on a large scale, and were used by banks, insurance companies, business, and government departments. They helped in tabulating data from censuses and large sample surveys. The desk calculator, referred to by Karl Pearson, consisted of circular discs with the digits 0 to 9 marked on them; these were mounted on a common horizontal axis and by turning a handle the machine could be made to add, subtract, multiply and divide. Desk calculators marked a revolution in the development of statistics, comparable to that due to computers in the 1960s.

Since early this century, routine statistical analysis of experimental data has been a part of agricultural and biological research; more recently it has become so in medical and industrial research. These analyses were very largely done on desk calculators, and Fisher (the man who has had most influence on the development of statistics) once said that most of his statistics had been learnt on the machine: without calculators, neither Karl Pearson nor Fisher could have achieved what they did. During these analyses of data, Fisher and his colleagues devised new statistical methods to deal with small samples of data and with data from designed experiments. Means and variances were of particular interest, and the central technique was the analysis of variance (see Chapter 12): this has as its basis a decomposition of a sum of squares. Calculating sums of squares was fairly straightforward on desk calculators, and became particularly easy when (some 40 years ago) the machines became powered by electric motors and incorporated extra keys to speed up the process of summing squares. Another development was to give machines extra storage registers, to hold intermediate steps in a calculation without so much need to copy numbers down on paper: this is often a source of errors. Large desk machines, used by research workers, could carry a considerable number of digits in a calculation; up to twenty in a sum of squares.

The first automatic electronic computers began to appear in the 1940s, and by 1960 many large companies, government departments, research institutes and universities had computer departments based on these machines. The important characteristic of the electronic computer is the very high speed at which it carries out calculations; hence it has to be automatic since no human brain could react quickly enough to keep control of a series of calculations. The computer must therefore be controlled by a program, or sequence of instructions, stored within it.

It was obvious that computers were very valuable in reducing large quantities of data from censuses and surveys, but most statisticians tend to be concerned with much smaller sets of data. Hence statisticians were slow to take up the use of electronic computers (although Rothamsted Experimental Station installed one in the early 1950s). They were put off by the lack of flexibility in analysis when using a computer; the early programs were mostly written by mathematicians and did not do exactly what the statisticians wanted, and the statisticians were on the whole unwilling to devote a lot of time to writing programs so long as their desk machines could do most of what they required. In a computer analysis, one loses direct contact with the data and cannot keep it, and the steps in an analysis, under close scrutiny as a statistician wishes to do—and can easily do when using a desk calculator. Although in theory the computer can carry out various checks during an analysis, early programs were not generally available to do this to the satisfaction of many statisticians; but perhaps the main criticism was of the 'one-shot' nature of the programs. The precise steps needed in a full statistical analysis can rarely be predicted completely at the outset, choice of later computations being very likely to depend on the results of earlier steps. In fact the chief users of statistical programs supplied by computer departments were, and most probably still are, non-statisticians. A research worker who has collected a set of data, very likely without consulting a statistician at the planning stage or even without any very precise plan at all, often sees his analysis as purely a piece of computing. He assumes that the program he chooses—or someone else chooses for him—will automatically give a satisfactory analysis of his data. Far too many analyses are done using programs which are not appropriate to the questions a set of data can answer. Often this is simply because a program is readily available; sometimes it is due to sheer lack of understanding or communication.

Some recent developments have made computers more acceptable to statisticians. Large computers often have multi-access interactive systems, with each of many users working at a terminal so that, provided that the system is not overloaded, he can act as if he alone is using the computer. Programs can be developed relatively quickly, and package programs (e.g. GLIM—General Linear Interactive Modelling) have been developed which allow the user to vary his planned analysis as results appear.

But what we see as a most important development for statistical analysis is the recent arrival of microcomputers. There has been a progressive and quite astounding miniaturisation of electronic computers. The early ones depended on valves; replacing these by transistors led to substantial reduction in size for the same level of power, and the more recent use of very small integrated circuit chips has made available desk computers which are of similar size to the old desk calculators, and are (allowing for inflation) also a similar price. But the computing power of a microcomputer is enormously greater than that of the desk machines; they can be programmed using a high-level language like a large mainframe computer, but they do not have the same problems of input and output, the complexity of which can waste much time and cause frustration to the ordinary user. So once again it becomes possible to keep an analysis under the user's control, to keep close to the original data while analysing it, and to watch the successive steps of the analysis for odd and interesting features.

We must not close this section without mentioning the pocket-size hand-held electronic calculators that first appeared in the 1970s. These have much to offer the occasional user of statistics: they allow quite complicated calculations to be done quickly and silently, without having to go to a computer room to punch cards or tape. Like the old electro-mechanical desk machines, they are of course prone to operator error, although unlike them the hand-held machines have a number of inbuilt programs, ranging from a mean-and-variance routine to (in more powerful versions) routines for regression and evaluating probability densities. It can be useful to have one available even when using a computer.

1.2   Programming a computer for statistics

Software does not have the same dramatic history of progress as does hardware. Nonetheless, the development from machine codes, which were the only languages available for early computers, to the high-level languages available today, such as Pascal, BASIC and FORTRAN, is a substantial achievement. But even if a high-level language is available, there is still the problem of how best to organise statistical computing.

Chambers (1980) points out three main approaches: (a) single programs; (b) statistical systems, or large package programs; (c) collections of sub-programs or algorithms. The single-program approach was adopted in developing the BMD Biomedical Computer Program at the University of California at Los Angeles. This was the first collection of statistical programs to be made generally available, the first manual being published in 1963; the programs have been very successful and continue, after many revisions, to be used. All the user needs to know is how to input and output data. However, with a collection of this type, combining programs may be awkward or impossible.

The second approach is to develop a statistical system. We can regard this as a very complex program allowing the user to carry out a range of statistical analyses by giving instructions in the special language of the system. Examples are SPSS (Statistical Package for the Social Sciences), developed in the USA, and GENSTAT (A General Statistical Program) and GLIM, both developed in Britain. These systems offer much to a regular user who has a full knowledge of the language of the system and appreciates the package's strengths and weaknesses; but the occasional user can find them quite incomprehensible! Usually such a system can be updated only by a complete revision; and since revisions are only likely to take place at considerable intervals of time, the system can easily fossilise.

The third approach is to develop a collection of statistical sub-programs or algorithms, and to combine these into programs as required. A convenient way of combining algorithms must be found, but if this can be done there is great flexibility in this method, which can be updated by introducing new algorithms. This book is essentially a collection of algorithms.

So far the computer has had a relatively small effect on basic statistics such as we chiefly consider in this book. Many of our programs are for procedures which can be done quite well on desk calculators, though of course if a satisfactory program is available the use of a computer reduces the tedium of this task. One vast area opened up by computers is that of simulation (see Chapters 7 and 9). Also the calculation of 'residuals', trivial on a computer (see Section 11.3) but prohibitively heavy on a desk calculator, helps to make possible fuller and more appropriate analyses of data. We shall also give (see Chapter 8) programs that allow the user to dispense with statistical tables and to calculate necessary values directly for himself. Graphical methods are an area in which the computer promises much but because we limit ourselves to what is available in standard Pascal our programs (see Chapter 5) can only hint at some of the possibilities.

The effect of computers is much more noticeable in advanced statistics. A common, and relatively simple, example is multiple regression, for which we give a program in Section 11.4 although we do not discuss it in detail. Regressions using a small number of variables and observations were often carried out on desk calculators, but the arithmetic became overwhelming if many variables were used. With a computer it is easy to include a very large number of variables; although unfortunately as the number increases so do the difficulties in interpreting the results, and in view of its wide use by non-statisticians one might confidently say that multiple regression has become the most misused technique of statistics. The whole field of multivariate analysis has been opened up through the advances in computers, but the techniques developed there are based on more complicated mathematics and require considerable computer storage space; so we have not extended the scope of this book to cover that field.

1.3   Representation of numbers in a computer

When using any calculating device, we must consider what errors will be introduced into calculations. In a computer, there is only a limited amount of space in which to store each number, and only the most significant digits in the number will be retained; the actual number of digits is governed by the word length of the particular machine. For example, consider calculating π^2 on a machine which will store six digits. The number $\pi = 3.1415926535 \ldots$ would be stored as 0.314159×10^1 (the decimal part is called the mantissa and the index of 10 the exponent). Multiplication of π by π would give, at best, $0.098695877281 \times 10^2$, which would be stored as 0.09869×10^2 and returned as 9.8696. As compared with the true value of π^2, this calculation gives an error in excess of 0.000004.

A second source of error is that computers operate on binary, not decimal, numbers. A number is represented in the computer by a sequence of switches, each of which may be *on* or *off*, in this way representing the binary digits 0 and 1. Thus each number used is converted from decimal to binary representation, and when the arithmetical operations are finished the resulting number is converted back to decimal. Whole numbers in the decimal system have exact binary equivalents, but most other real numbers do not, even though they may be decimal fractions. For example, $0.1 = \frac{1}{10}$ has the infinite binary representation 0.0 0011 0011 0011 . . . and cannot therefore be stored exactly in the computer. The effect of this is easily seen by running a short program based on the following code:

```
program add(input,output);

var i : integer;
    t : real;

begin
    t := 0;
    for i := 1 to 500 do
    begin
        t := t + 0.1;
        writeln(t)
    end
end.
```

Since most computers convert a decimal number into a decimal fraction with an exponent, as a first step in calculation, in the way that we wrote π as 0.314159×10^1 above, there is almost certainly an extra error involved in converting the mantissa to a binary number. Similar errors occur with most arithmetic operations carried out in a program, and there is a real danger that the error will accumulate to the extent that several significant digits will be in error. In the worst cases, accumulation of round-off error and loss of significant digits can invalidate a calculation completely.

Consider calculating $f(x) = \dfrac{p}{1 - x^2}$ for $p = 3.0$ and $x = 0.99900050$. Because of rounding, x is taken as 0.99900000 (an error in the seventh decimal place). The result obtained is 1500.7504, but it should be 1501.5007: an error in the *seventh* decimal place has produced an error in the *fourth* significant figure of $f(x)$. The critical step here is subtracting x^2 from 1; these numbers are very similar in magnitude, and this leads to significant digits being lost in the divisor. The detailed analysis of such errors is beyond the scope of this book and we recommend the interested reader to consult texts on numerical analysis, e.g. those by Johnson and Riess (1982) and Wilkinson (1963).

1.4   Interactive and batch computing

There are two common methods of using a computer. In both the user communicates directly with the machine using a video-terminal, a combination of keyboard and visual display unit. In the first method, instructions are typed at the keyboard and passed to the computer; the results are seen, almost instantaneously, on the visual display unit. This may be called **interactive computing**. The second method is to prepare, using a keyboard, a program file and, if necessary, data files. These are stored within the computer, usually on disks, to be run at a convenient time. The output from such a program is again usually stored on disk, for printing or for display on a screen, at the user's convenience. This is called **batch processing**.

In the early days of computing, batch processing was the common method; the program and data 'files' were prepared on punched cards, or paper tape, away from the computer. But the turn-round time was at best hours and might be days, so that developing a program, with the many errors that involves, took a long time. Also although some alternatives could be built into a program, the user could not step in after seeing the results of the early part of the program and say what was to be done in the later part. (This second facility is particularly valuable in statistical analysis. Only as one finds out more about the data can the best form of analysis be decided.) Most large computers now have many terminals attached to them so that many users may share the time of the computer and use it in an interactive way (and, of course, microcomputers with a single operator are used interactively).

Thus interactive development of programs, and the use of interactive programs, are usually the best and most convenient ways of working. However, when the sets of data are large, or several analyses have to be carried out, then it may be less wasteful of machine resources, and cause less delay to other users, to use batch processing.

1.5   Exercises

1   Carry out the following calculations with a, b, c and d as if you are a computer that works in two-significant-digit arithmetic. Thus evaluate each component arithmetic operation exactly and then ignore all digits in the result except the first two significant digits.

$$a = 0.62 \times 10^4, \ b = 0.23 \times 10^{-3}, \ c = 0.16 \times 10^1, \ d = 0.58 \times 10^4.$$

(i) $[a.b]/c$, (ii) $a.[b/c]$, (iii) $b.[a-d]$, (iv) $b.a - b.d$.
The brackets [ ] denote which operation is done first. The exact results are: (i) and (ii), 0.89125×10^0; (iii) and (iv), 0.9200×10^1.

2   Write a program which finds the smaller root of

$$0.001 \, x^2 + 100.1 \, x + 10000 = 0$$

by using:

(a) the usual formula;
(b) the relation that smaller root = (product of roots)/(larger. root); the product of roots is equal to the constant term in the equation divided by the coefficient of x^2, and the larger root is found from the usual formula.

3   The integral

$$I_n = \frac{1}{e} \int_0^1 x^n e^x \, dx \qquad \text{where } n = 1, 2, \dots$$

can be calculated from the following recurrence relation (derived by integrating by parts):

$$I_n = 1 - n I_{n-1} \qquad n \geq 1$$

and $I_0 = 1 - \dfrac{1}{e}.$

From the form of the integrand it is clear that $I_n \to 0$ as $n \to \infty$. Investigate what happens if the integral is evaluated by this method, using successively cruder estimates for the value of e; to 9 figures, $e = 2.71828183$.

2 Some general principles

2.1 Computation of a mean

We shall begin our consideration of statistical program writing by looking at what is probably the most common calculation in statistics: finding the arithmetic mean of a set of data. For a small set of data we do not need a computer to do this. However, if we consider how best to program this simple calculation we can identify some useful general principles of programming without being distracted by the detail of a complex central calculation.

We give first Program 2.1, which calculates a mean. It does what is required but it is not a good program. In the next section, we give an improved version: Program 2.2, *Mean*. We shall find it useful to refer to these two programs as examples.

```
program exercise(input,output);

var k,u : integer;
    a : real;
    b : array[1..100] of real;

begin
readln(k);
for u := 1 to k do
readln(b[u]); a := 0;
for u := 1 to k do
a := a + b[u]; writeln(a/k)
end.
```

Program 2.1 *Exercise*

A useful reference on the general approach to programming and on programming style is Schneider and Bruell (1981).

2.2 Properties of a good program

The prime aim of a good program is to carry out its task successfully and accurately; appropriate and reliable input must be operated upon to produce clear, understandable output. In addition, if people are to use the program well, the listed program must have a clear structure and be easy to understand. These properties are the ones that we shall stress.

We have not mentioned efficiency yet. Often this is used to mean machine efficiency only, and it is sometimes gained at the expense of human inconvenience; in other words, it may cause human inefficiency. For example, milliseconds of computing time might be saved by an intricate piece of coding, but the consequence may be that a slight modification of the program takes hours of work, because the programmer has difficulty in understanding the code. Yet machine efficiency is important. Provided human efficiency has been ensured, we want programs to be economical of storage and to be quick. The best way to achieve this is to use a good algorithm, i.e. a good general approach to the problem. We have used the best algorithms that we can find. Another way to improve machine efficiency is to tighten up the coding, e.g. if an array element is used a number of times then instead of looking it up each time, which is a slow operation, look it up just once and store it in a simple variable for future use. The reader will find that he can sometimes tighten up our code. If he wishes to use a program a very large number of times this may be worth doing. We have often refrained from doing so if we thought it might make a program less understandable.

```
program Mean(input,output);

(* prints the mean of n elements of an array x *)

var i : integer; (* label of observation *)
    Mean : real;
    n : integer; (* number of observations *)
    Sum : real; (* temporary store when adding observations *)
    x : array[1..100] of real; (* observations *)
begin
    write('State number of observations  ');
    readln(n);
    writeln;
    writeln('Input Observations');
    for i := 1 to n do
        readln(x[i]);

    Sum := 0;
    for i := 1 to n do
        Sum := Sum + x[i];
    Mean := Sum/n;

    writeln('Number of observations = ',n);
    writeln('Mean = ',Mean)
end. (* of program Mean *)
```

Program 2.2 *Mean*

2.3   Making a program understandable

A program should not only fulfil its purpose accurately and efficiently, it should also be easily understandable by another user. (That other user may be yourself in a few months' time – you may well have entirely forgotten or, worse, remember imprecisely what your program will and will not do.) We recommend a number of practices to assist you in making a program intelligible. We wish to refer to procedures and functions as well as to programs so, in this section, we shall use the term 'algorithm' to refer to all three. In the rest of the book we refer to procedures and functions as 'algorithms' but keep 'program' as a separate word.

(a)   *Choose meaningful variable names.* This, more than any other of our suggestions, makes a program understandable. The variable names in Program 2.1, for example '*u*' and '*a*', convey no information, whereas the names '*Mean*' and '*Sum*', in Program 2.2, immediately indicate the nature of the variables. Names should be descriptive and unambiguous. Standard Pascal is very helpful in this respect since it allows names to be of unlimited length (though some implementations demand that a given number of initial characters, often 8, provide a unique name). Long names can be clumsy and unwieldy (e.g. *ResidualSumofSquares*) so often a balance between clarity and conciseness must be found (for this example, we might use *SSResidual*). Some names include a number of words; there can be no spaces in a variable name so, to pick out the words, we begin each word with a capital letter and print the rest of the word in lower case letters (e.g. *LowerEndPoint*). In a similar way, names consisting of only one word are written with a capital letter followed by lower case letters (e.g. *Residual*). Some statistical quantities are commonly referred to by single letters and we have used some of these. Thus we use '*n*' for the total number of observations and '*x*' for an array of observations, so that individual observations are written '$x[1], x[2], \ldots$'. We also use '*i, j, k*' for labels, counters or subscripts.

(b) *Give the algorithm a meaningful title.* Program 2.2 has the title '*Mean*', which indicates what the program is concerned with, whereas Program 2.1 has the nondescript and unhelpful title '*Exercise*'. If the name of a procedure is given in a program, this calls the procedure. Therefore a program reads well if the name of each procedure used in the program describes what that procedure does. Looking ahead to Program 2.3 for example, we see procedures named '*FindMean*' and '*OutputMean*'. Titles of functions should specify what is calculated. Thus we use a function '*NormalProb*' to calculate the probability that a standard Normal variable takes a value less than some specified value.

(c) *Annotate the algorithm.* Comments may be placed freely throughout a Pascal program provided they are included within curly brackets '{ ... }' or within round brackets and asterisks '(∗ ... ∗)'. Examples of the following recommendations may be seen in Program 2.2.

Immediately following the title there should be a prologue describing what the algorithm does. In procedures and functions the prologue should include a description of the parameters of the algorithm. Variables whose purpose is not clear from their names should, on declaration, be followed by a brief description. If an algorithm divides naturally into distinct sections it may help to give a brief description, or title, at the beginning of each section. 'End' statements should usually be annotated to link up with the beginning of the appropriate complex statement.

(d) *When declaring variables, list the names in alphabetical, or other meaningful, order.* We adopt the convention, when declaring variables, of giving the list of variables in alphabetical order, with one variable per line and with comment where necessary. (Closely related variables can be put on the same line.) As may be seen in Program 2.2, the list gives a good indication of the function of the program, and is also a convenient reference.

(e) *Write out the algorithm to exhibit its structure.* The most important device is indentation: degree of indentation should indicate level of computation. In Program 2.2 there is a loop:

for $i: = 1$ to *n* do

  $Sum: = Sum + x[i]$;

The action of summing takes place inside the loop and may be considered as being at a deeper level of computation than the control statement 'for . . . do'. A similar device may be used with the other types of loop, with 'case' statements and with 'if . . then . . else' statements; examples will be found throughout the book. The idea of level of computation is a useful general guide but the precise interpretation can become a matter of personal opinion. For example, the best rule for placing 'begin' and 'end' is not obvious. They act as brackets for complex statements and in this book we place them at the same level as the control statement, e.g.

for $i: = 1$ to n do

begin

   $Sum1: = Sum1 + x[i];$

   $Sum2: = Sum2 + abs(x[i])$

end;

but they could well be indented, i.e.

for $i: = 1$ to n do

   begin

   $Sum1: = Sum1 + x[i];$

   $Sum2: = Sum2 + abs(x[i])$

   end;

A line space between distinct sections of the program, as in Program 2.2, can make the program more readable.

2.4   Program structure

A good principle to follow in order to make a program easy to understand and easy to modify is to construct it in distinct sections. It is often possible to develop and test sections of the program independently and hence isolate errors more easily. If changes are to be made it is much easier to see what needs to be done if one's efforts can be concentrated on a small section instead of the entire program. Pascal is constructed to encourage this modular approach by making available procedures and functions.

   Program 2.2 is quite satisfactory for the simple job it has to do. But to exhibit the general approach of this book, in which we present a collection of algorithms that may be put together to form a program, we shall divide the program into procedures. This makes it over-elaborate for what it has to do but allows us to illustrate the use of procedures.

   The active section of the program, after the declarations, may be divided into three parts with distinct tasks: (1) input the data, (2) calculate the mean, (3) output the mean. All statistical programs involving data (and most other programs) have the same three-part structure which we may summarise:

   read
   calculate
   write.

We would recommend, as a general rule, that each part be written as one or more procedures.

We have rewritten Program 2.2, *Mean*, using procedures and have produced Program 2.3, *MeanRevised*. We have also taken the opportunity to elaborate the program a little. By looking at the main body of Program 2.3, at the end of the listing, we see immediately the simple three-part structure of the program.

```
program MeanRevised(input,output);

(* prints the mean of n elements of an array x *)

const Dp = 2; (* number of decimal places in output of mean  *)
      Field = 8; (* width of field in which mean is output    *)
      MaxSampleSize = 100; (* maximum number of observations *)

type Units = 1..MaxSampleSize;
     DataVector = array[Units] of real;

var Mean : real;
    n : Units; (* number of observations *)
    x : DataVector; (* observations *)


procedure InputData(var x : DataVector; var n : Units);

(* input of a number n and an array of n elements *)

var i : integer; (* label of observation *)

begin
    write('State number of observations   ');
    readln(n);
    writeln;
    writeln('Input Observations separated by spaces');
    writeln('Check data on each line before pressing RETURN');
    for i := 1 to n do
        read(x[i]);
    writeln
end; (* of InputData *)


procedure FindMean(x : DataVector;  n : Units;
                   var Mean : real);

(* calculates mean of n elements of array x *)

var i : integer; (* label *)
    Sum : real; (* temporary store when adding observations *)

begin
    Sum := 0;
    for i := 1 to n do
        Sum := Sum + x[i];
    Mean := Sum/n
end; (* of FindMean *)
```

```
procedure OutputMean(Mean : real; n : integer);

begin
    writeln('Number of observations = ',n);
    writeln('Mean = ',Mean)
end; (* of OutputMean *)

begin (* main body of program *)
    InputData(x,n);
    FindMean(x,n,Mean);
    OutputMean(Mean,n)
end.
```

Program 2.3 *MeanRevised*

2.4.1   Parameters of algorithms

A Pascal function calculates a single value. Many functions need to be supplied with arguments, or parameters, to determine the precise value calculated. For example, we might declare a function *Factorial*(n), that calculates $n(n-1) \ldots 2.1$, where n is a positive integer. In a program we might call *Factorial* (*4*), or *Factorial* (*i*), where *i* is the name of an integer variable in the calling program. Pascal terminology is to say that both '*4*' and '*i*' are **value parameters**. We may think of them as input values for the function sub-program.

Pascal procedures can be much more complex sub-programs than functions, and the situation with respect to parameters is more complicated. Two types of parameters are recognised: **value parameters** and **variable parameters**. (Variable parameters are distinguished in the parameter list by being preceded by **var**.) A simple way to think of them, that covers most of the cases we are concerned with, is to consider the value parameters as inputs to the procedure and the variable parameters as outputs. For example, a procedure to calculate the mean and median of three observations named *a*, *b*, *c* would have as value parameters *a*, *b*, *c* and as variable parameters *Mean* and *Median*, to choose appropriate names. The first line of the procedure declaration might be:

procedure *FindMeanMedian* (*a*, *b*, *c*: real; var *Mean*, *Median*: real);

An example of its use, in a program, would be the statement:

FindMeanMedian (*6, 7, 3, xMean, xMedian*);

where *xMean* and *xMedian* are the names of variables in the calling program.

Another example would be:

FindMeanMedian (*p, q, r, xMean, xMedian*);

where *p, q, r* are also variables in the calling program but it is only their values, at the time of calling, that are passed to the procedure.

Our simple way of classifying the parameters breaks down if a variable is used as an input to a procedure but we wish to use an updated value of the variable in the calling program. Any variable whose value is changed by a procedure must be classified as a variable parameter.

Array parameters require a special mention. If a variable is passed as a value parameter, a local copy of the variable is made for use in the procedure. For an array of large dimension this is very demanding of storage space and may create difficulties when trying to run the program. Therefore some programmers adopt a general rule of always passing arrays as variable parameters. We do not do this, but declare array parameters according to the rules we gave above, since this makes clear the status of the array. But if the reader is concerned about storage in a particular application, he may wish to change our declaration of an array parameter in the parameter list of a procedure.

2.4.2 Array declarations

The size of an array that we wish to use in a procedure will depend on the number of data. We would therefore like to be able to change the size of an array in a parameter list. We know of two approaches to this problem. One is to declare, in the main program, an array of sufficiently large dimension that it will exceed all the array sizes that will be required in the procedures used in a particular series of runs. (One might imagine the analysis of several sets of data, each with a different number of observations). We thus declare a large array size and use as much of it as we want; this is the approach we adopt. Another approach is to use conformant arrays; this has conveniences if a software library is being developed but introduces an extra complication for the programmer. We note also that our approach is acceptable to any Pascal standard whereas conformant arrays are part of the later standard (BS 6192) and so are not universally implemented (see Wilson and Addyman, 1982). Conversion to conformant arrays is not a difficult task.

2.4.3 Type declarations

We have introduced, in Program 2.3, some type declarations that we shall use throughout the book. They are assumed in many of the procedures. We define an index type *Units* which consists of the integers from *1* to *MaxSampleSize*, where *MaxSampleSize* is a constant. A further type is *DataVector* which is a one-dimensional array, over *Units*, of real numbers. We declare in Program 2.3, the variable *n*, which represents the number of observations and is of type *Units*, and the variable *x*, which represents the set of observations $x[1]$, $x[2]$, . . . and is of type *DataVector*. The variable names *x* and *n* are often used in this sense in the book. Occasionally we require a type similar to *DataVector* but with integer instead of real elements; we call this type *DataList* (page xi).

In the later chapters of the book we make use of a two-dimensional data structure to represent *n* samples of each of *p* variates. We first declare an index type *Variates* which consists of the integers from *1* to *MaxVar*, where *MaxVar* is a constant. Then we declare a type *DataMatrix* which is an array over *Variates* of *DataVector*. The fundamental elements of variables of type *DataMatrix* are of type real. A similar structure whose fundamental elements are of type integer is declared to be of type *DataTable* (see page xi).

From recorded data we may derive frequency tables. We therefore declare one- and two-dimensional data types for these tables. We declare an index type *Classes* (= *1*. .*MaxClass*) and then a type *FreqList* to be an array over *Classes* of integer (page xi). A similar type with real elements is called *FreqVector*. For two dimensions the index types are *Rows* (= *1*. .*MaxRow*) and *Cols* (= *1*. .*MaxCol*), to produce a type

FreqTable which is an array over *Rows* and *Cols* of integer. A similar type with real elements is called *FreqMatrix*.

Declarations of **records** will be discussed in Chapter 12.

2.5 Ensuring reliable input

When typing in data there is an appreciable chance that you may make a mistake. You might type in 34 instead of 43, 17.3 instead of 1.73, or even type in a letter instead of a digit. The important point is that data must always be checked.

It is not appropriate to give a full discussion of checking input in this introductory chapter (we do so in Chapter 13). At this stage we recommend simple methods that are adequate for small data sets (up to 30 observations, say). Building in adequate safeguards for larger data sets quickly leads to a substantial algorithm.

For small data sets we recommend that the input be checked by reading over a copy of the data. In batch processing, using cards, the data cards will be put at the bottom of the pack, after the program. The data values may be printed on each data card and checked; cards with incorrect values may be replaced. Checking, by reading over the data, is not so convenient in interactive computing (at least when using simple 'read' or 'readln' instructions) so more care is needed. When using the 'readln' instruction as in Program 2.1, it would be necessary to check each observation as it appeared on the monitor screen, immediately after it had been typed in. If the number were correct, RETURN could be pressed and the number accepted; if the number were incorrect, it could be modified before acceptance. This is a tedious method, so we have made a slight modification when producing the *InputData* procedure in Program 2.2. Input of observations is done using the 'read' instruction. This allows data to be entered using 'free format', i.e. numbers are regarded as distinct provided they are separated by one or more spaces, or by a RETURN. This allows a line of data, containing say five observations, to be entered. The data can be read over and incorrect values modified, using the simple line editing facilities of the teletype, before accepting the values by typing RETURN. Checking is easier if the data are arranged in a regular manner with, for example, the same number of observations in each row and numbers arranged underneath each other. Using this method the check that the correct number of observations has been presented is easily made. One technical point in the coding of the *InputData* procedure should be mentioned. A 'readln' instruction is inserted after the input of the data using the 'read' instruction. This ensures that the pointer in the input file ends up at an appropriate place.

Any call for input in an interactive program should be accompanied by a printed instruction to the user. This has not been done in Program 2.1 and the user might well not appreciate what the computer requires. This weakness of the program has been remedied in Program 2.2 and, even more fully, in Program 2.3.

2.6 Making output clear and understandable

Output values should always be clearly labelled. This is yet another failing of our example 'bad' program, Program 2.1. There is labelling of output in Programs 2.2 and 2.3 but it is very simple. It consists only of 'Mean = ', together with the number of observations. In any real-life application, a title for the data and/or the name of the variate should be printed. We do not do it in Programs 2.2 and 2.3 because we would need to introduce procedures for the input and output of string variables; as introductory examples we think the programs are long enough. (Input of string variables is dealt with in Chapter 13.)

A typical output of the mean value in Program 2.2 would be 9.93000000000000E + 01, which is read as '9.93 times 10 to the power 1', or 99.3. The default action of Pascal is to print real numbers in floating-point notation. The value is not clear or easy to read. In the procedure *OutputMean*, in Program 2.3, we have used the facilities to (1) state the width of the field within which the value is printed, and (2) state the number of decimal places. The typical value above would appear as 99.30, which is much clearer. It is convenient to declare the values chosen for field width ('*Field*') and number of decimal places ('*Dp*') as constants at the beginning of the program listing, so that they may be changed easily. More discussion of presentation of output will be found in Chapter 4.

2.7 Exercises

1 Run Program 2.2, with the following set of data, and check that your result agrees with that given.

  72, 58, 60, 68, 59, 72, 85, 73, 57, 76. (Mean = 68.0).

2 Modify Program 2.2 so that it can deal with several sets of data, and print out the mean for each set.

3 Data on the performance of a new machine are collected over a period of several days; each day several records of a quantity X are taken. At the end of each day, it is required to find (i) *DayMean*, the mean of that day's observations, and (ii) *OverallMean*, the mean of all the data collected so far. Write a program which does this *without* working through every item of previous days' data, but updates the previous *OverallMean* instead.

4 Modify Program 2.2 so that it calculates and prints the geometric mean as well as the arithmetic mean. [The geometric mean of a set of numbers $x_1, x_2, \ldots, x_n$ equals $\sqrt[n]{(x_1 x_2 \ldots x_n)}$; it may be calculated by forming the product and taking the nth root, although here there is a danger that the product may overflow the register, or by using the fact that

$$\mathrm{Ln}(GeometricMean) = \frac{1}{n} \sum_{i=1}^{n} \ln x_i.$$

Try both methods of calculation. As a test set of data for x, use the values 1.0 E2, 1.0 E3, 1.0 E4, 1.0 E5, 1.0 E6, which have geometric mean 1.0 E4 = 10^4.]

5 It may be inconvenient to have to input the number of observations before each set of data. One way of overcoming this, which can be useful if applied carefully, is to put a **marker**, which is a very large number like 999999, at the end of each data set and to incorporate in the input program a test of when the end of the set is reached. (A disadvantage of this is that 'outliers', or members of the data set which are large, although not as large as the marker, might also need to be traced for study.) A negative integer can be used as a marker if the data themselves are certain to be positive.

Modify Program 2.2 to contain a marker and an appropriate test.

3 Sorting and ranking

3.1 Introduction

A number of useful descriptive statistics and statistical tests depend on relative magnitudes, or order properties, of observations. For example, we may wish to pick out the maximum and minimum observations in a set, or the middle observation (the median) in a set. Sometimes it may even be useful to list a whole population or sample in order of size, from smallest to largest. Several statistical tests use the **ranks**, i.e. the position numbers of the observations when the whole set has been put in order.

The fundamental procedure used in determining most of these quantities is **sorting** the data. Sorting is an extremely important computer method, and much has been written about it. We have space to touch only the borders of the subject; thorough discussions are given in Knuth (1973) and Lorin (1975). Wirth (1976) discusses sorting with particular reference to Pascal.

3.2 The Exchange sort

The most efficient methods of sorting are not easy to understand. So we shall give first a method that is easy to understand and simple to program; it is satisfactory to use for small collections of numbers—say less than 30—but very slow for large collections.

We illustrate the method with a small set of data, given in the order

10, 3, 14, 17, 2, 11.

Begin with the observation on the extreme left, 10; compare it with the observations to the right. If any of these are smaller, as are 3 and 2, choose the smallest, 2, and exchange it with 10:

2, 3, 14, 17, 10, 11.

Now consider the second observation from the left, 3; compare it with the observations to its right. None is smaller, so no exchange is made. In the third position is 14; we find that the smallest number to its right is 10, and so we exchange these two:

2, 3, 10, 17, 14, 11.

It is easy to see that by following this procedure we end with the sequence:

2, 3, 10, 11, 14, 17.

If we assume the data to be sorted are the first n elements of an array x, the method may be described in the following pseudo-code:

for i: = 1 to *n* − 1 *do*
 begin 'assign index of the lowest of *x*[*i*], . . . , *x*[*n*] to *low*';
   'exchange *x*[*i*] and *x*[*low*] if they are different' *end.*

The complete description is given in Algorithm 3.1, *ExchangeSort*.

```
procedure ExchangeSort(var x : DataVector; n : Units );

(* sorts the first n elements of array x in ascending order  *)

var    i,j : integer; (* loop counters *)
       Low : integer; (* index of smallest of elements being  *)
                      (* considered                           *)
       xValue : real; (* temporary store when making exchange *)

begin
    for i := 1 to n-1 do
    begin
        Low := i;
        for j := i+1 to n do
            if x[j] <= x[Low] then Low := j;
        if Low <> i
        then begin
            xValue := x[i];
            x[i] := x[Low];
            x[Low] := xValue
        end; (* of exchange *)
    end; (* of pass through data *)
end; (* of ExchangeSort *)
```

Algorithm 3.1 *ExchangeSort*

3.3 Shellsort

With *n* items of data, the number of comparisons made in *ExchangeSort* increases approximately as n^2. The time taken to sort a small set of data is quite acceptable, and the simplicity of the algorithm makes it a good choice. But if it is often required to sort large sets of data, *ExchangeSort* may be irritatingly slow for routine use. It becomes worthwhile to use a more efficient method, even at the expense of more complex programming.

 We give next a method, first proposed by D L Shell, that has become generally known as Shellsort; it is also called the 'merge-exchange sort'. For large sets of data it is appreciably quicker than *ExchangeSort*: the time taken increases approximately as *n*ln*n*. We illustrate the method by sorting the twelve observations:

   7, 9, 2, 11, 4, 1, 3, 10, 5, 6, 2, 8.

Begin by comparing all pairs of observations which are *n* div 2 apart; exchange the observations in a pair if they are not in the correct order already. For our set of twelve observations, we list on the same line the pairs to be compared and underline the observations which are exchanged.

First pass: <u>7</u>   <u>3</u>
                9   10
                2   5
              <u>11</u>   <u>6</u>
                <u>4</u>   <u>2</u>
                ‾‾   ‾‾
                1   8.

Having made the exchanges, the data are next merged:

   3, 9, 2, 6, 2, 1, 7, 10, 5, 11, 4, 8.

The gap between observations to be compared is now halved (working if necessary to the next integer below); in our example the gap becomes 3. We list again on the same line the sets to be compared, and underline those observations which are moved.

Second Pass: 3   6   7   11
                <u>9</u>   <u>2</u>   <u>10</u>   <u>4</u>
                <u>2</u>   <u>1</u>   5   8.

Let us consider the exchanges and comparisons involving the numbers on the second line (these operations take place between operations on other numbers). First, the number '9' is compared with the number '2' and an exchange made:

   2   9   10   4.

Then '9' is compared with '10', and these numbers remain as they are. Next '10' is compared with '4' and these two numbers are exchanged:

   2   9   4   10.

When it makes sense to do so, a primary comparison is followed by a number of secondary comparisons to ensure that all the numbers that have been inspected on a line are in order. Thus the primary comparison, and exchange, of '10' with '4' is followed by the secondary comparison, and exchange, of '4' with '9':

   2   4   9   10;

it is followed also by the secondary comparison of '4' with '2', although this leads to no exchange. The observations are again merged:

   3, 2, 1, 6, 4, 2, 7, 9, 5, 11, 10, 8.

The gap between observations to be compared is next reduced to 3 div 2, i.e. 1, which will be the final gap length every time Shellsort is used.

Third Pass: 3 2 1 6 4 2 7 9 5 11 10 8.

In this example, all observations move in the final pass, as a consequence either of a primary comparison or of a secondary comparison. But because of the earlier long-distance comparisons they move little in the final pass.

Final Merge: 1, 2, 2, 3, 4, 5, 6, 7, 8, 9, 10, 11.

   The method is programmed in Algorithm 3.2, *ShellSort*. Note that, in the inner 'repeat' loop, observation *j* is compared with observation *j* + *Gap*. If the condition fails, which means that the two observations are already in increasing order, the program passes out of the loop and the next pair of observations is considered. Otherwise, observations *j* and *j* + *Gap* are exchanged and as a secondary step observation *j* − *Gap* is compared with observation *j*; the consequence of this comparison may again be either to pass out of the loop or to make an exchange and stay in the loop.

```
procedure ShellSort(var x : DataVector; n : Units );

(* sorts the first n elements of array x in ascending order *)

var Gap : integer; (* gap between observations being compared *)
    i,j : integer; (* loop counters *)
    Nextj : integer; (* j + gap *)
    xValue : real; (* temporary store when making exchange *)

begin
    Gap := n;
    repeat
        Gap := Gap div 2;
        if Gap > 0
        then begin
            for i := 1 to n-Gap do
            begin
                j := i;
                while j >= 1 do
                begin
                    Nextj := j+Gap;
                    if x[j] > x[Nextj]
                    then begin
                        xValue := x[j];
                        x[j] := x[Nextj];
                        x[Nextj] := xValue
                    end (* of exchange *)
                    else j := 0;
                    j := j-Gap
                end (* of while loop *)
            end (* of comparison for given i *)
        end; (* of pass with given gap size *)
    until Gap = 0
end; (* of ShellSort *)
```

Algorithm 3.2 *ShellSort*

3.4   A comparison of sorting methods

In order to show how much the different sorting methods can vary in speed, we have run a number of sorting algorithms on a VAX11/780 machine using the same sets of observations; we have recorded the times taken and other relevant data. Besides *ExchangeSort* and *ShellSort*, we have included in these trials three further methods. One of these is 'Bubblesort' (or 'Ripplesort'), which is widely used but which is not a method that we would recommend; the algorithm is described in question 7 of

Exercises 3.9. The other two methods are very rapid but more complicated. We describe them briefly and give algorithms (3.3 and 3.4).

3.4.1 Quicksort

In Algorithm 3.3, *QuickSort*, the complete list of observations to be sorted is partitioned into two sub-lists, such that the numbers in one sub-list are all less than or equal to a given number, called the **fence**, while the numbers in the other sub-list are all greater than the fence. This partitioning process is repeated on the sub-lists, using fences chosen from each sub-list, and is continued until sub-sub- . . . -lists are produced which each contain one observation; the complete array is then in order. The procedure *QuickSort* uses a subsidiary procedure *Partition* which acts recursively.

```
procedure QuickSort(var x : DataVector; n : Units );

(* sorts the first n elements of array x in ascending order *)

    procedure Partition(LeftIndex,RightIndex : integer );
    var   Fence : real; (* dividing element for partition *)
          i,j : integer; (* indexes of array elements *)
          xValue : real; (* temporary store for exchanges *)

    begin
        i := LeftIndex;
        j := RightIndex;
        Fence := x[(LeftIndex+RightIndex) div 2];
        repeat
            while x[i] < Fence do
                i := i+1;
            while Fence < x[j] do
                j := j-1;
            if i <= j
            then begin
                xValue := x[i];
                x[i] := x[j];
                x[j] := xValue;
                i := i+1;
                j := j-1
            end (* of exchange of values *)
        until i > j;
        if LeftIndex < j then Partition(LeftIndex,j);
        if i < RightIndex then Partition(i,RightIndex)
    end; (* of partition *)

begin
    Partition(1,n)
end; (* of QuickSort *)
```

Algorithm 3.3 *QuickSort*

3.4.2 Tree sort

All the sorting algorithms described so far have involved exchanges within the data array. Since each exchange involves three instructions, a method that could avoid exchanges would seem to have a great advantage. In *TreeSort* (Algorithm 3.4), each item

```
procedure TreeSort(var x : DataVector; n : Units );

(* sorts the elements of array x in ascending order *)

type   Pointer = ^Element;
       Element = record
                     Value : real;
                     Left : Pointer;
                     Right : Pointer
                 end;

var    Add : Pointer;
       i : integer; (* counter *)
       Top : Pointer;

    procedure MakeTree(var Ptr : Pointer );
    begin
        if Ptr = nil
        then   Ptr := Add
        else
            if Add^.Value > Ptr^.Value
            then MakeTree(Ptr^.Right)
            else MakeTree(Ptr^.Left)
    end; (* of MakeTree *)

    procedure StripTree(Ptr : Pointer );
    begin
        if Ptr <> nil
        then begin
            StripTree(Ptr^.Left);
            x[i] := Ptr^.Value;
            i := i+1;
            StripTree(Ptr^.Right)
        end
    end; (* of StripTree *)
begin
    Top := nil;
    for i := 1 to n do
    begin
        new(Add);
        Add^.Value := x[i];
        Add^.Left := nil;
        Add^.Right := nil;
        MakeTree(Top)
    end; (* completes tree *)
    i :=1;
    StripTree(Top)
end; (* of TreeSort *)
```

Algorithm 3.4 *TreeSort*

in the array is considered in turn and, by means of pointers, is located on a branching tree structure in such a way that it is possible by following the pointers to read out the numbers in order. This algorithm makes use of the record data structure and of pointers. The two subsidiary procedures *MakeTree* and *StripTree* are used recursively.

All the sorts may be modified to sort lists of words, or records which may contain several words and numbers. Moving these string variables around is a slow operation, and *TreeSort* has the advantage that it does not move the variables. It can also be adapted to sort several variables simultaneously. For example, a set of records might consist, for each member of a group of people, of the person's name, age and address. By including extra arrays of pointers, these variables could be sorted with a single pass through the data.

3.4.3   Sorting times

The sorting times for the five methods we have introduced are given in Table 3.1, which also contains information on the numbers of comparisons and exchanges. There are no exchanges in *TreeSort*, and the moves in *QuickSort* cannot be classified as simple exchanges, so there is no figure for 'number of exchanges' given for either of these sorts. The figures given in the tables are averages taken over 100 blocks of n random numbers. For each sorting technique, the random number generator was initialised identically, so that the blocks of numbers being compared by the various techniques really were the same. Comparisons and exchanges were counted by inserting incremental counters just before each comparison and exchange. The times of execution were found in separate trials using the CLOCK function, which is a feature of VAX-Pascal and is not in

Table 3.1

Time (milliseconds) to sort n random numbers

n	100	500	1000
Bubblesort	109	2445	9820
Exchange sort	86	1914	7628
Shellsort	30	176	398
Tree sort	63	358	774
Quicksort	27	137	305

Number of comparisons when sorting n random numbes

n	100	500	1000
Bubblesort	4830	124285	498834
Exchange sort	4950	124750	499500
Shellsort	841	6462	15503
Tree sort	1222	7621	16722
Quicksort	1087	6613	14006

Number of exchanges when sorting n random numbers

n	100	500	1000
Bubblesort	2208	61579	251131
Exchange sort	98	496	990
Shellsort	376	2942	7990

standard Pascal. The instruction T1: = CLOCK was inserted as the first executable instruction of the procedure and T2: = CLOCK was inserted immediately before the final END statement of the procedure. TIME: = T2 − T1 thus gave the time spent in the procedure.

We see from Table 3.1 that both *ShellSort* and *QuickSort* do well although *QuickSort* has the edge in speed. For applications involving only small sets of data, *ExchangeSort* might be preferred for its simplicity, but with more than 100 observations neither it nor Bubblesort can be seriously considered. As we have said earlier, we see nothing to recommend Bubblesort; it involves about the same number of comparisons as *ExchangeSort* but makes many more exchanges.

TreeSort is expensive on storage; it requires $3n$ locations when sorting n numbers, whereas *QuickSort* needs only $n + \log_2 n$. (*ExchangeSort* and *ShellSort* each require n locations.) *TreeSort* is useful when there is additional information associated with the items being sorted. In all the sorting methods we have mentioned except *TreeSort*, items are moved around while being sorted so that the link with additional information is lost, whereas after *TreeSort* the additional information is still accessible.

3.5   Ranking

Given an array of data x, it may be useful to know the rank of each element in it. These values are calculated in Algorithm 3.5, *FindRanks*. The algorithm counts how many observations are smaller than any given observation $x[i]$; this count is labelled *NoLess*.

```
procedure FindRanks(x : DataVector; n : Units;
                    var Rank : DataVector);

(* generates array called Rank whose i-th element is the *)
(* rank of x[i], the smallest element has rank 1.        *)

var    i,j : integer; (* loop counters *)
       NoEqual : integer; (* counts number of observations *)
                          (* equal to a given observation  *)
       NoLess : integer; (* counts number of observations *)
                         (* less than a given observation *)
       xValue : real; (* observation being considered *)

begin
    for i := 1 to n do
    begin
        NoLess := 0;
        NoEqual := 0;
        xValue := x[i];
        for j := 1 to n do
            if x[j] = xValue
                then NoEqual := NoEqual+1
                else if x[j] < xValue
                        then NoLess := NoLess+1;
        Rank[i] := NoLess+(NoEqual+1)/2
    end (* of ranking x[i] *)
end; (* of FindRanks *)
```

Algorithm 3.5 *FindRanks*

If all the observations are different, the rank of $x[i]$ is $NoLess + 1$. Note that the smallest observation has rank 1.

A modification to this simple process is needed if any of the observation-values are repeated: a **tie**. Thus in the ordered set of numbers

10, 11, 13, 14, 14, 17, 19,

there are two 14's, having ranks 4 and 5. In such a case it is the custom to give each occurrence of the repeated number the mean rank, here 4.5; then 17 will be given the rank 6. In Algorithm 3.5, *NoEqual* counts how many times each number occurs in the data set. The rank for each occurrence of a number that is repeated is then given as $Rank[i] = NoLess + (NoEqual + 1)/2$.

This ranking algorithm, 3.5, is satisfactory for small sets of data and has the great virtue of simplicity, but for large data sets it is very slow. It is possible to augment a sorting program to find the ranks of the observations. Algorithm 3.6, *RankAndSort*, does this. In the first part of the algorithm the observations are sorted, and their movements recorded in the array *StartPlace*. In the second part, ranks are assigned taking into account possible ties.

In a similar way we can augment other sort algorithms in which data are moved in an array; this adds little to the sort time.

```
procedure RankAndSort(var x,Rank : DataVector; n : Units);

(* generates an array called Rank whose i-th element is the *)
(* rank of the input element x(i); the smallest number has  *)
(* rank 1. On exit the array x contains the data sorted in   *)
(* ascending order.  The dimension of array x must be at     *)
(* least n + 1 where n is the number of data in the array x  *)

var  i,j : integer; (* loop counters *)
     Low : integer; (* index of smallest of elements being    *)
                    (* considered *)
     NoEqual : integer; (* counts number of observations      *)
                        (* equal to a given observation        *)
     RankValue : real; (* rank of observations being  *)
                       (* considered *)
     StartPlace : array[1..50] of integer; (* StartPlace[i] *)
                  (* is the initial position of the x value *)
                  (* currently in x(i) *)
     TotalRank : real; (* sum of ranks of observations *)
                       (* being considered *)
     xValue : real; (* an element of x *)
     Value : integer; (* temporary store for exchanges *)
begin
    for i := 1 to n do
        StartPlace[i] := i;
    (* sort and record moves *)
    for i := 1 to n-1 do
    begin
        Low := i;
        for j := i+1 to n do
            if x[j] <= x[Low] then Low := j;
```

```
        if Low <> i
        then begin
            xValue := x[i];
            x[i] := x[Low];
            x[Low] := xValue;
            Value := StartPlace[i];
            StartPlace[i] := StartPlace[Low];
            StartPlace[Low] := Value
        end; (* of exchange of values *)
    end; (* of sort *)

    (* rank observations taking account of ties *)
    NoEqual := 1;
    TotalRank := 1;
    xValue := x[1];
    x[n+1] := xValue - 1;
    for i := 1 to n do
    begin
        if x[i+1] = xValue
        then begin
            NoEqual := NoEqual + 1;
            TotalRank := TotalRank + i + 1
        end (* of count of equal values *)
        else begin
            RankValue := TotalRank/NoEqual;
            for j := i + 1 - NoEqual to i do
                Rank[StartPlace[j]] := RankValue;
            xValue := x[i+1];
            NoEqual :=1;
            TotalRank := i + 1
        end; (* of rank assignment *)
    end; (* of ranking *)
end; (* of RankAndSort *)
```

Algorithm 3.6   *RankAndSort*

3.6   Maximum and minimum values

These values can of course be read off from a data array once it has been sorted; but they can also be found by a single pass through unsorted data, as in Algorithm 3.7, *FindMaxMin*. The current upper bound is called *Maximum* and the lower bound *Minimum*. Initially both *Maximum* and *Minimum* are set equal to the first observation $x[1]$. The values of *Maximum* and *Minimum* are modified if necessary as each observation is considered.

3.7   Quantiles

The quantiles are a useful class of descriptive statistics. They are observations (or notional observations) which divide, in specified proportions, the total frequency of a set of observations. The most commonly-used quantile is the **median**: this is the observation that divides the total frequency in half—or, in other words, it is the middle observation. If n (the total frequency) is an odd number, the median is the $(n + 1)/2$th observation in rank order: for, assuming that all the observations are distinct, there will

```
procedure FindMaxMin(x : DataVector; n : Units;
                     var Maximum,Minimum : real );

(* finds maximum and minimum values of an array x *)

var   i : integer; (* loop counter *)
      xValue : real; (* element of x being considered *)

begin
    Maximum := x[1];
    Minimum := Maximum;
    for i := 1 to n do
    begin
        xValue := x[i];
        if xValue > Maximum
            then Maximum := xValue
            else if xValue < Minimum
                then  Minimum := xValue
    end (* of loop indexed by i *)
end; (* of FindMaxMin *)
```

Algorithm 3.7 *FindMaxMin*

be $(n-1)/2$ observations smaller than the median and the same number larger than it. If n is even, there is no 'middle' observation, and so we put the median midway between the $n/2$th and the $(n/2+1)$th observations, since this value will have $n/2$ observations below it and $n/2$ above. This illustrates the common practice when calculating quantiles: if a dividing point does not fall on an observation, we estimate the quantile as the mean of the two observations between which it falls. When using Algorithm 3.8, p should be restricted to the range 1 to $q-1$ (inclusive).

The quartiles divide the total frequency into quarters, the deciles divide it into tenths and the percentiles divide it into hundredths. Algorithm 3.8, *Quantile*, is a function for calculating the p-th q-tile from a data array that has already been sorted. Thus $p = 3$, $q = 4$ gives the third quartile; $p = 2$, $q = 10$ gives the second decile; $p = 1$, $q = 2$ gives the median. The quantile is calculated from both ends of the set of data (i.e. beginning with the largest as well as beginning with the smallest); if these two calculated values coincide the quantile coincides with one of the observations. If the values do not coincide, we estimate the quantile to be half the sum of these two values.

```
function Quantile(x:DataVector; n:Units; p,q:integer):real;

(* finds p-th q-tile of data,in ascending order, in array x *)

var i1,i2 : integer; (* ranks of quantile counting from left *)
                     (* and right *)

begin
    i1 := (p*n) div q + 1;
    i2 := n-((q-p)*n div q);
    Quantile := (x[i1] + x[i2])/2
end; (* of Quantile *)
```

Algorithm 3.8 *Quantile*

3.8   Test data for Algorithms

Algorithm 3.1, *ExchangeSort*. (i) Input: $n = 6$, $x = (2, 7, 1, 8, 2, 8)$.
Output: $x = (1, 2, 2, 7, 8, 8)$. (ii) See question 1 of Exercises 3.9, below.
Algorithms 3.2, *ShellSort*; 3.3, *QuickSort*; 3.4, *TreeSort*. Same as for Algorithm 3.1,
ExchangeSort.
Algorithm 3.5, *FindRanks*. Input: $n = 6$, $x = (2, 7, 1, 8, 2, 8)$. Output: $Rank = (2.5, 4, 1,$
$5.5, 2.5, 5.5)$.
Algorithm 3.6, *RankAndSort*. Input: as for Algorithm 3.1, *ExchangeSort*.
Output: x and $Rank$ as in Algorithms 3.1, 3.5.
Algorithm 3.7, *FindMaxMin*. Input: as for Algorithm 3.5, *FindRanks*.
Output: $Maximum = 8$, $Minimum = 1$.
Algorithm 3.8, *Quantile*. Input: $n = 6$, $x = (1, 2, 2, 7, 8, 8)$. $p = 1$, $q = 2$.
Output: $Quantile\ (x, 6, 1, 2) = 4.5$.

3.9   Exercises

(*Many exercises, in this and later chapters, make use of random numbers. There may be a random number function available on your computer; alternatively Algorithm 7.2, Random, may be used.*)

1   Write a program to sort 25 random numbers using Algorithm 3.1, *ExchangeSort*.
Output the result.

2   Vary the size of the list of random numbers in question 1 and record the times taken.

3   Put in increasing order the first 21 digits of π:

3.141 592 653 589 793 238 46

4   Write a flow diagram for *ExchangeSort*, Algorithm 3.1.

5   The time t taken in sorting methods should depend on the size n of the list of numbers sorted. Find a relation of the form $t = kn^c$ for *ExchangeSort* and for *QuickSort* (see Table 3.1).

6   Modify *ExchangeSort* (Algorithm 3.1) or *ShellSort* (Algorithm 3.2) so that it will put a list of words into alphabetical order. Words may be handled by defining a variable type, called *String* say, which is a packed array of characters. The statement:

*Word*1 > *Word*2

is true if the string-type variable *Word*1 is after *Word*2 in alphabetical order.
   Test your program by putting the days of the week in alphabetical order.

7   *Bubblesort*. Comparisons are made between the numbers as in Exchange Sort, but an exchange is made each time a number is found that is out of order. Thus at the end of the first pass the largest number has moved to the end of the list. The next pass, which ends at the last position but one, moves the second largest number to its correct place. An algorithm for Bubblesort, using the same input variables as Algorithm 3.1, *ExchangeSort*, is as follows.

```
procedure BubbleSort(var x : DataVector; n : Units);

var i,j,k : integer; (* loop counters *)
    Test : boolean;
    Temp : real; (* temporary storage during exchange *)

begin
    i := 0;
    repeat
        i := i + 1;
        Test := false;
        for j := 1 to n - i do
        begin
            k := j + 1;
            if x[j] > x[k] then
            begin
                Temp := x[k];
                x[k] := x[j];
                x[j] := Temp;
                Test := true
            end
        end;
        if Test = false then i := n - 1;
    until i = n - 1;
end; (* of BubbleSort *)
```

How does this algorithm differ from *ExchangeSort*, and why is it slower?

8 Modify *QuickSort* (Algorithm 3.3) so that the fence is chosen at random from the values $x[LeftIndex] .. x[RightIndex]$. Compare the speed of sorting of Algorithm 3.3 with the speed of the modified Algorithm when the numbers to be sorted are in: (a) increasing order, (b) decreasing order, (c) random order.

9 A useful method of scanning data to detect 'wild' observations, errors in typing, misplaced decimal points, etc. is to calculate the maximum, minimum and mean values of the data set. It is useful to identify which items in the list are the maximum and the minimum. Develop a program to carry this out using the following procedure.
1) Read the number of data *n*; read *n* data into an array *x*.
2) Find the mean of the data in *x* (see Program 2.2).
3) Find the maximum and minimum values of the data (see Algorithm 3.4) and to which items these belong.
4) Print out number of data, minimum value, mean and maximum value, with the numbers labelled (e.g. MEAN 6.32) and necessary identifications made.

Run the program with the following set of 10 observations: 13.7, 16.2, 162, 12.9, 11.0, 21.5, 12.6, 14.5, 11.7, 12.1.

10† Write a program to:
1) calculate the mean *m* of a set of *n* data in an array *x*;
2) calculate the number n_1 of data less than *m* and the number n_2 greater than *m*;
3) print the mean *m* and an index $(n_1 - n_2)/n$.

[This index measures how close the mean is to the median, being 0 if the mean and median coincide and having a large absolute value if they are far apart.]

Run the program using the following two sets of data.
a) Weekly wages, in £, of ten workers (including the manager) in a factory: 96, 95, 160, 94, 95, 93, 98, 92, 97, 91.
b) Heights of ten plants in a greenhouse, recorded in mm: 31, 27, 43, 35, 37, 33, 34, 36, 39, 36.

11   Amend *TreeSort* (Algorithm 3.4) so that it is able to sort records on two different criteria with a single pass; for example, put names in alphabetical order and ages in increasing order.

12   A procedure for an 'insert sort' is given below. Describe briefly the method that is being used. Compare the speed of this *InsertSort* with the speeds of the other sort procedures introduced in this chapter.

```
procedure InsertSort(var x : DataVector; n : Units);

type   pointer = ^element;
       element = record
                      Value : real;
                      Link : pointer
                 end;

var    Ptr,Last,Top,Newptr : pointer;
       Found : boolean;
       i : integer;

begin
new(Top);
Top^.Value := 0.0;
Top^.Link := nil;
for i := 1 to n do
    begin
    Found := false;
    Ptr := Top;
    repeat
       Last := Ptr;
       Ptr := Ptr^.Link;
       'if Ptr = nil then
           Found := true
         else
           Found := Ptr^.Value > x[i]
    until Found;
    new(Newptr);
    Newptr^.Value := x[i];
    Newptr^.Link := Ptr;
    Last^.Link := Newptr
    end;
Top := Top^.Link;
for i := 1 to n do
    begin
    x[i] := Top^.Value;
    Top := Top^.Link
    end
end; (* of InsertSort *)
```

13† Write a program to calculate:
i) the **range** of a set of data, which is the difference between the maximum and minimum observations;
ii) the **interquartile range**, which is the difference between the upper and lower quartiles (one-half of this, the 'semi-interquartile range', is often used instead).

Incorporate this program into one which calculates, and prints the values of, the mean, median, range and semi-interquartile range of a set of observations. Make sure the output is labelled clearly.

Run the program on the following set of data: 87, 67, 98, 57, 74, 100, 83, 60, 99, 88, 54, 72, 78, 75, 93.

4 Inspection and summary of data using tables

4.1 Introduction

Tables and diagrams are the main aids available to help us inspect and present data. In this chapter we discuss how the computer can be used to compile tables; in Chapter 5 we shall deal with graphical methods.

Our first concern must always be that the correct data have been put into the computer. Input data must always be checked. We have already discussed this briefly in Section 2.5; we give a full discussion in Chapter 13.

The essence of a good table is layout; we therefore also discuss output. It may surprise the reader to learn that input and output are usually much more awkward to program than are calculations. We often cannot produce as neat and intelligible an output as we would like; but, with a little care, we can produce a substantial improvement on unconsidered, unplanned output.

4.2 Input of an array

We extract the procedure *InputData* from Program 2.3 and repeat it here as Algorithm 4.1 for convenient reference. We stress again that this algorithm is suitable only for small sets of data. A more elaborate algorithm is required for bigger sets of data if adequate checks are to be incorporated.

```
procedure InputData(var x : DataVector; var n : Units);

(* input of a number n and of an array x of n elements *)

var i : integer; (* label of observation *)

begin
    write('State number of observations ');
    readln(n);
    writeln('Input observations separated by spaces');
    writeln('Check data on each line before pressing RETURN');
    for i := 1 to n do
        read(x[i]);
    readln
end; (* of InputData *)
```

Algorithm 4.1 *InputData*

4.3 A six-number summary

The data we shall be considering in this chapter will usually be a sample from some population that we are interested in; occasionally, however, it may be a whole population. For example, we may be given the prices of packets of the same brand of coffee in a sample of shops, or we may have the complete rainfall figures for each day in July in a town in Great Britain.

The first, very simple, summary table we shall look at is based on the ideas of Tukey (1977), though we have slightly modified his proposals. Suppose we have a sample consisting of the heights (in cm) of nine 13-year-old boys:

157, 147, 164, 152, 143, 156, 170, 159, 160.

The procedure we use is to put the observations in increasing order and then pick out (1) the minimum, or lowest, value; (2) the first quartile, *Quartile*[1]; (3) the median, or *Quartile*[2]; (4) *Quartile*[3]; (5) the maximum value, or upper bound. It is convenient to call the minimum *Quartile*[0] and the maximum *Quartile*[4].
These quantities have been underlined in the ranked observations:

143, 147, 152, 156, 157, 159, 160, 164, 170.

It is reasonable to take these values, which are spread out evenly through the data, as being representative of the sample. We present them in a table which also includes the sample size (or, if we had a complete population, we should give instead the total number of observations).

Sample size	9	
Median	157	
Quartiles	152	160
Extremes	143	170

Note that, with some sample sizes, the median and quartiles will fall between observations (see Section 3.7).

In producing the data for the table, we may use algorithms that we have already constructed. The assembled program is shown in Program 4.2, *SixNumberSummary*. We have used a shorthand for the algorithms that have already been introduced, to save copying them out completely. The algorithm *PrintSixNoSummary* is a new one and is given in full. Few statements are required in the main program section apart from those calling the algorithms.

Program 4.2, *SixNumberSummary*, is a complete program with input, calculation and output sections. We shall discuss the output of the six-number summary in Section 4.4.

4.4 Output of the six-number summary

We put data in a table to make clear, to ourselves or to others, the relations between numbers; the way numbers are spaced on a page brings out their interrelations. We shall consider in detail the output of the six-number summary of Section 4.3, as an example of instructing a computer to give us the output we require. The reader will quickly find (or has, very likely, done so already!) that when we output anything more complex than

```
program SixNumberSummary(input,output);

(* calculates and prints out the six number summary of *)
(* a set of data.                                       *)

const MaxSampleSize = 30;

type   Units = 1.. MaxSampleSize;
       DataVector = array[Units] of real;
       QuartileArray = array[0..4] of real;

var    n : Units; (* sample size *)
       p,q : integer; (* parameters in p-th q-tile *)
       Quartile : QuartileArray;
       x : DataVector; (* data *)


procedure InputData(var x : DataVector; var n : Units);

        (*** Algorithm 4.1 ***)


procedure ShellSort(var x : DataVector; n : Units);

        (*** Algorithm 3.2 ***)


function Quantile(x: DataVector; n : Units; p,q:integer):real;

        (*** Algorithm 3.8 ***)

procedure PrintSixNoSummary;

begin
    writeln('Sample size ',n:16);
    writeln('Median ',Quartile[2]:21:2);
    writeln('Quartiles ',Quartile[1]:12:2,Quartile[3]:12:2);
    writeln('Extremes ',Quartile[0]:13:2,Quartile[4]:12:2)
end; (* of PrintSixNoSummary *)


begin  (* main program *)
    InputData(x,n);
    ShellSort(x,n);

    (* pick out extremes *)
    Quartile[0] := x[1];
    Quartile[4] := x[n];

    (* calculate quartiles *)
    q := 4;
    for p := 1 to 3 do
        Quartile[p] := Quantile(x,n,p,q);
    PrintSixNoSummary
end. (* of program SixNumberSummary *)
```

Program 4.2 *SixNumberSummary*

a simple number on a computer we must take care, or we shall find that the numbers pop up in completely unexpected parts of the screen or page.

The naïve way to output the six-number summary is with the coding:

writeln ('Sample size', *n*);
writeln ('Median', *Quartile*[2]);
writeln ('Quartiles', *Quartile*[1], *Quartile*[3]);
writeln ('Extremes', *Quartile*[0], *Quartile*[4]);

We tried this on a VAX computer and obtained the following output:

Sample size	9	
Median	1.57000E + 02	
Quartiles	1.52000E + 02	1.60000E + 02
Extremes	1.43000E + 02	1.70000E + 02

This does not make a good table. It is possible to do much better.

Let us look more closely at the output we have obtained and see what rules the computer has been following. When printing the value of the integer variable *n*, in the first line, the computer has put 7 spaces after the words 'Sample size' and then printed the value of *n* in the next 3 spaces. The total space allowed is called the **field** and the number of spaces in the field is called the **field-width**. The value of the integer is printed at the right-hand end of the field, using as many spaces as necessary. The field-width of 10 is the default setting (i.e. what is done if no specific instruction is given) for the computer we were using. We can specify the field-width we require by inserting the value, in the 'write' instruction, as a positive integer following the variable, and separated from the variable by a colon. We give a revised coding of the output instructions below and have replaced '*n*' by '*n*: 16'. A field-width of 16 allows the value of the sample size to fit better into the complete output.

The default setting for the printing of integer variables gives us something that is readable. It is the printing of real variables that is very confusing in our example. The default printing, on our computer, gives a field-width of 12 and fills the complete field with the number in floating-point form. The combination of many digits and the indices make the numbers very difficult to read. Fortunately, Pascal gives us the option of printing real variables in fixed-point form. To do this we follow the name of the variable, in the 'write' instruction, with (1) the field-width and (2) the number of decimal places required; the parameters are separated from the variable name, and each other, by colons. Examples can be seen in this revised coding of the ouput:

writeln ('Sample size', *n*:16);
writeln ('Median', *Quartile*[2]:21:2);
writeln ('Quartiles', *Quartile*[1]:12:2, *Quartile*[3]:12:2);
writeln ('Extremes', *Quartile*[0]:13:2, *Quartile*[4]:12:2);

When this is used we obtain:

Sample size	9	
Median	157.00	
Quartiles	152.00	160.00
Extremes	143.00	170.00

This method requires us to count spaces to ensure that the alignment is satisfactory, but gives a much better output. For the data of our example, a value of zero for the number of decimal places would give a neater output; we have chosen two decimal places as a value that would be generally satisfactory.

4.5   Frequency table for discrete variate with repeated values

Suppose we wish to form a frequency table from a set of observations which are the number of letters a man received each weekday over a long period:

$$1, 0, 3, 1, 2, 0, 0, 3, 1, 1, \ldots .$$

A count of the frequency of each value of the variate may be obtained using a simple algorithm, *CountFrequencies*, 4.3. This algorithm is for use when we have a variate taking integer values from 0 up to some upper bound. The maximum value, *LargestX*, which might be obtained using *FindMaxMin* (Algorithm 3.7), must be stated; or, alternatively, any convenient upper bound may be given.

```
procedure CountFrequencies(x : DataList; n : Units;
                           LargestX : integer;
                           var Frequency : FreqList);

(* produces a frequency table from the n items of data in   *)
(* the array of integers x.  LargestX must be not less than *)
(* any element of x.                                         *)

var i : integer; (* values of elements in x *)
    j : integer; (* particular element of x *)

begin
    for i := 0 to LargestX do
        Frequency[i] := 0;
    for i := 1 to n do
    begin
        j := x[i];
        Frequency[i] := Frequency[i] + 1;
    end
end; (* of CountFrequencies *)
```

Algorithm 4.3 *CountFrequencies*

4.6   Frequency table with data grouped in classes

Data in which the values of observations are repeated are not too common. More often we need to group data into classes in order to obtain a frequency table (*ABC*, Section 1.2.2). The choice of class-intervals, so as to produce a convenient table, is not a simple decision, and we shall consider in Section 4.6.1 how best to make the choice. For the moment let us assume that the lower limit l, the upper limit u and the class-width d have been chosen. The intervals are given by

$$l \leq x < l+d, l+d \leq x < l+2d, \ldots, l+(c-1)d \leq x \leq u$$

where *c* is the number of classes; its calculation is described later.

The frequency table is calculated in Algorithm 4.4, *FormFreqTable*. The algorithm is a straightforward generalisation of Algorithm 4.3, *CountFrequencies*. The rank number of the class in which *x*[*i*] is located is determined. These values, which must be integers,

```
procedure FormFreqTable(x : DataVector; n : Units;
                        LowerLimit,UpperLimit,
                        ClassWidth : real;
                        var Frequency : FreqList);

(* produces a frequency table, using classes of size   *)
(* ClassWidth from the n items of data in the array of real *)
(* numbers x.  All the elements of x must lie in the range  *)
(* LowerLimit to UpperLimit.                            *)

var i : integer; (* loop counter *)
    ClassRank : integer;
    NoOfClasses : integer;
    ScaledValue : real;

begin
  NoOfClasses:=trunc(UpperLimit-LowerLimit)/ClassWidth)+1;
  for i := 1 to NoOfClasses do
      Frequency[i] := 0;
  for i := 1 to n do
  begin
      ScaledValue:=(x[i]-LowerLimit)/ClassWidth;
      ClassRank := trunc(ScaledValue) + 1;
      Frequency[ClassRank] := Frequency[ClassRank] + 1
  end
end; (* of FormFreqTable *)
```
Algorithm 4.4 *FormFreqTable*

```
procedure OutputFreqTable(Frequency : FreqList;
                          LowerLimit,UpperLimit,
                          ClassWidth : real;
                          NoOfClasses : integer;
                          Field,Dp : integer);

(* prints the frequency table produced by FormFreqTable *)

var i : integer; (* loop counter *)
    LeftEndPoint,RightEndPoint : real; (* class boundaries *)

begin
    for i := 1 to NoOfClasses-1 do
    begin
        RightEndPoint := LowerLimit + i*Classwidth;
        LeftEndPoint := RightEndPoint - ClassWidth;
        writeln(LeftEndPoint:Field:Dp,' <= X < ',
                RightEndPoint:Field:Dp,' ',Frequency[i]:Field)
    end;
    writeln(RightEndPoint:Field:Dp,' <= X <= ',UpperLimit:
            Field:Dp,' ',Frequency[NoOfClasses]:Field)
end; (* of OutputFreqTable *)
```
Algorithm 4.5 *OutputFreqTable*

are counted in the same way as in *CountFrequencies*. The output of the results is carried out using Algorithm 4.5, *OutputFreqTable*.

4.6.1   Choice of class-intervals

We look at two examples of data sets, in order to see what points have to be considered.

(a)   *Examination marks* (%). The data might run: 61, 39, 48, 51, 22, . . . with maximum 83 and minimum 22. We know without looking at the data that the possible values range from 0 to 100. A class-width of 10 is an obvious natural choice, leading us to the convenient class-intervals 0–9, 10–19, etc. Setting the lower limit to zero, the upper limit to 100, and the class-width to 10 would produce these intervals; in the output the intervals would be described as

$$0 < = X < 10, \ 10 < = X < 20, \text{ etc.}$$

Since the maximum and minimum values in our set of data are 83 and 22, we shall obtain zero frequencies in the class-intervals 0–9, 10–19 and 90–99. There is no disadvantage in this, and the intervals suggested are probably the best choice for these data.

An automatic method of choosing class-intervals would be to set the minimum value (here 22) at the lower limit and the maximum value (83) at the upper limit. If the number of classes were fixed at 12 then the class-width would be $(83-22)/12 = 5.0833333$. The intervals would be

$$22 < = X < 27.0833333, \ 27.0833333 < = X < 32.1666666, \text{ etc.}$$

This example shows at once that any automatic method, unless backed up by a substantial program, might give very awkward class-intervals. We recommend therefore that the choice of intervals should be a matter of personal judgement, using perhaps the maximum and minimum values in the data set as guides to intelligent choice.

Choice of the lower limit *l* and the class-width *d* fixes the sequence of class-intervals, but the computer must also be told when to stop. The *lower* limits of the class-intervals begin at *l* and continue, as far as possible, in the sequence $l+d, \ l+2d, \ldots$, provided that they do not exceed the stated upper limit *u*. There is no necessity for *u* to be an integral number of class-widths from *l*. The final class-interval will in fact always be shorter than the others, and it is probably best to aim to make it contain no observations. In the case of the class-intervals we have recommended for the examination-marks data, the final 'interval' will consist of the single point $x = 100$. The final interval has been made to end at *u* in case larger values are ridiculous—as for example 103% would be in the examination marks.

(b)   *Lengths of cuckoo eggs* (mm). The data quoted in *ABC*, Section 1.2.2, run: 22.5, 20.1, 23.3, 22.9, 23.1, . . . with maximum 25.0 and minimum 19.6. As in the manual production of a frequency table, discussed in *ABC*, an approximate class-width is given by (maximum − minimum)/10, i.e. 5.4/10 or 0.54. We thus see that a convenient choice of the class-width *d* is 0.5; and we might choose *l* to be 19.0 and *u* to be 25.5.

4.7 Two-way table from classification data

We take the opportunity here to use non-numeric data which are stored as character variables in the computer. We consider a survey in which two questions are asked, each of which has a fixed number of possible answers.

To illustrate the method, suppose that we are collecting information to determine whether left-handedness is more common in women than in men. For each person in our sample we shall therefore record whether they are male (M) or female (F), and we shall ask whether they are left-handed. Of course, some people use both left and right hands for various tasks, so we must choose an unambiguous definition of 'left-handed'. For simplicity, ask only which hand they write with, and record L for left, R for right.

Our aim is to produce a 2×2 table in this example (more generally it will be a table with a chosen number r of rows and c of columns); in the table the rows will refer to one classification factor (sex in our example) and the columns to another factor (handedness). An extra row and an extra column carry the totals:

Freq	L	R	Total
M	CellTotal[1, 1]	CellTotal[1, 2]	RowTotal[1]
F	CellTotal[2, 1]	CellTotal[2, 2]	RowTotal[2]
Total	ColTotal[1]	ColTotal[2]	n

Program 4.6, *ContingencyTable*, has been written to carry out our particular example. But the central procedure *FormTwoWayTable* is a general procedure for forming a $r \times c$ table. The remainder of the program can easily be modified for other examples.

```
program ContingencyTable(input,output);

(* produces a two-way contingency table summarising data on *)
(* sex and handedness                                        *)
const MaxSampleSize = 100;
      NoOfRows = 2;
      NoOfCols = 2;

type Class = 1..2;
     ColRange = 1..NoOfCols;
     RowRange = 1..NoOfRows;
     Units = 1..MaxSampleSize;
     List = array[Units] of Class;
     FreqTable = array[RowRange,ColRange] of integer;
     FreqList = array[RowRange] of integer;

var CellTotal : FreqTable;
    ColTotal : FreqList;
    Error : boolean;
    HandCode : char;
    i,j : integer;
    n : Units;
    RowName : packed array[1..10] of char;
    RowTotal : FreqList;
    SexCode : char;
    x,y : List;
```

```
procedure FormTwoWayTable(x,y : List; n : Units;
                          var CellTotal : FreqTable;
                          var RowTotal : FreqList;
                          var ColTotal : FreqList);

(* forms a two-way table of frequencies given row and column *)
(* classification of each unit.                              *)

var i,j : integer; (* current classes *)
    k : integer; (* loop counter *)

begin
    for i := 1 to NoOfRows do
    begin
        RowTotal[i] := 0;
        for j := 1 to NoOfCols do
            CellTotal[i,j] := 0;
    end;
    for j := 1 to NoOfCols do
        ColTotal[j] := 0;
    for k := 1 to n do
    begin
        i := x[k];
        j := y[k];
        CellTotal[i,j] := CellTotal[i,j] + 1;
        RowTotal[i] := RowTotal[i] + 1;
        ColTotal[j] := ColTotal[j] + 1
    end
end;


begin (* main program *)

    (* input values *)
    write('State number of units  ');readln(n);
    writeln;writeln('For each unit');
    writeln(' (1) Type sex code, M or F, followed by RETURN,');
    writeln(' (2) Type hand code, L or R, followed by RETURN');
    writeln;

    i := 1;
    while i <= n do
    begin
        repeat
            Error := false;
            writeln(' ');
            writeln('    Unit  ',i:4);
            write('Sex Code   ');readln(SexCode);
            write('Hand Code  ');readln(HandCode);
            if SexCode in ['M','m','F','f']
            then case SexCode of
                    'M','m' : x[i] := 1;
                    'F','f' : x[i] := 2
                 end
            else Error := true;
            if HandCode in ['L','l','R','r']
            then case HandCode of
                    'L','l' : y[i] := 1;
                    'R','r' : y[i] := 2
```

```
          end
          else Error := true;
          if Error
          then writeln('Re-enter: M/F, RETURN, L/R, RETURN');
      until not Error;
      i := i + 1
   end; (* of while *)

   FormTwoWayTable(x,y,n,CellTotal,RowTotal,ColTotal);

   (* write table *)
   RowName[1] := 'M'; RowName[2] := 'F';
   writeln;writeln;writeln;
   writeln('freq':6,'L':6,'R':6,'Total':8);writeln;
   for i := 1 to NoOfRows do
   begin
       write (RowName[i]:6);
       for j := 1 to NoOfCols do
           write(CellTotal[i,j]:6);
       writeln(RowTotal[i]:6)
   end;
   write('Total':6);
   for j := 1 to NoOfCols do
       write(ColTotal[j]:6);
   writeln(n:6)
end.
```

Program 4.6 *ContingencyTable*

4.8 Test data for Algorithms

Algorithm 4.1, *InputData.* Input as required with $n = 6$. Array to be (11, 21, 31, 59, 79, 99). Make mistakes on entry and correct them.

Program 4.2, *SixNumberSummary.* Input: $n = 9$. Array to be (41, 31, 21, 17, 27, 37, 44, 33, 22). Output: $n = 9$, $Quartile[2] = 31$, $Quartile[1] = 22$, $Quartile[3] = 37$, $Quartile[0] = 17$, $Quartile[4] = 44$.

Algorithm 4.3, *CountFrequencies.* Input: $n = 8$, $LargestX = 6$, $x = (2, 3, 4, 5, 6, 5, 4, 3)$. Output: *Frequency* = (0, 0, 1, 2, 2, 2, 1).

Algorithm 4.4, *FormFreqTable.* Input: $n = 12$, $LowerLimit = 2.0$, $UpperLimit = 10.0$, $ClassWidth = 2.0$, $x = (2.3, 3.4, 4.5, 5.6, 6.7, 7.8, 8.9, 9.8, 8.7, 7.6, 6.5, 5.4)$. Output: *Frequency* = (2, 3, 4, 3, 0).

Algorithm 4.5, *OutputFreqTable.* Output the data from Algorithm 4.4 above.

Program 4.6, *Contingency Table.* Input: $n = 10$. Pairs of data (M, L), (F, L), (M, L), (F, R), (F, L), (M, R), (F, L), (M, R), (M, L), (M, L). Output: $CellTotal(1, 1) = 4$, $CellTotal(1, 2) = 2$, $CellTotal(2, 1) = 3$, $CellTotal(2, 2) = 1$.

4.9 Exercises

1 Write a program which generates 100 random digits and prints out a frequency table of the digits. We would expect each digit to occur with approximately equal frequency.

2†   Modify the program in question 1 so that it (1) reads *n*, and then reads *n* data into *x*; (2) calculates the maximum value from the data. Run your program with the following data:

(a) 10, 3, 7, 2, 4   5, 6, 12, 6, 3   8, 3, 3, 4, 4   9, 2, 4, 2, 7   3, 7, 4, 2, 4 (these are the numbers of letters in the first 25 words of this chapter);

(b) the first significant digit and, separately, the third significant digit in atomic masses (which can be looked up in a table of atomic masses);

(c) the first significant digit in lists of geographical data such as lengths of rivers, heights of mountains or populations of towns.

3†   Write a program which
1) inputs data (use Algorithm 4.1, *InputData*);
2) calculates the maximum and minimum values from the data (use Algorithm 3.7, *FindMaxMin*);
3) makes a frequency count of the data (use Algorithm 4.4, *FormFreqTable*);
4) prints out the results.

   Try out your program using some of the data given or described in question 2.

4   Write algorithms to carry out the following operations on a set of *n* data in an array *x*:

(a) given the number of decimal places required, print the data in a column with all decimal points in line;

(b) print the data in a column 'left-justified', i.e. having all first digits in line.

5†   Add the algorithms *FormFreqTable* (4.4) and *OutputFreqTable* (4.5) to Program 4.2, *SixNumberSummary*, to produce a program which:
1) prints the six-number summary;
2) asks the user to input the limits and class-width for the frequency table;
3) prints the frequency table;
4) returns to the input of limits and class-width to allow these to be changed and a new table produced.

6†   Write an algorithm to produce a two-dimensional frequency table. From *n* pairs of data (x_k, y_k) stored in arrays *x*, *y* produce a two-way frequency table *CellTotal*:*CellTotal* $[i, j]$ is to equal the number of pairs such that x_k falls in the *i*th class of the variable X and y_k falls in the *j*th class of the variable Y. [The algorithm is a generalisation of Algorithm 4.4, *FormFreqTable*. If the reader has difficulty in doing this, he will find such an algorithm in the first part of *PlotScatter* (5.4).]

5 Inspection and summary of data using graphical methods

5.1 Introduction

The relationships between variables in a set of data can be shown up much more vividly by graphical methods than by numbers alone. But it can be very tedious to plot graphs using only pencil and paper, and often we are too lazy to do so! Our excuse is removed, however, if we have a computer to do the work for us.

Many machines now have quite elaborate facilities for graphical work. We discuss these briefly in Section 5.6, but we shall first concentrate on what is available in standard Pascal.

On a computer screen we are restricted to a coarse matrix of positions. Common sizes are 40×24 positions (40 columns $\times 24$ rows) and 80×24 positions; in each position we may print a character (letter, figure or other symbol). Suppose we wish to draw a scatter diagram. First we must make decisions like those required when drawing the plot on graph paper: we must fix the position and length of the axes, and we must choose appropriate scales on the axes. This is a complex set of decisions. There do exist programs for carrying out this process which ensure, for example, that scales are convenient and that plotting on the available grid produces minimum misplacement of the points. Unfortunately, one of these programs would take up more space than we can afford. So we shall adopt a simple, crude approach which, while far from perfect, will give results that are broadly satisfactory.

Within the screen area available, we choose a space, which we call the **window**; inside this the plot is framed. We fix window width and window height, each of which is measured as a number of character spaces. For each variable we choose upper and lower limits for its values ($xMin$ and $xMax$ respectively for variable x, let us say); these may actually be chosen to coincide with the maximum and minimum values of the observations in a set of data. We then evaluate $ScaleFactor = WindowWidth/(xMax - xMin)$, giving the number of character spaces corresponding to each unit of the variable. Thus a point which is d units from the y-axis is $d * ScaleFactor$ character spaces away from the y-axis on the screen. More precisely, it will actually be $trunc$ $(d * ScaleFactor + 1)$ spaces from the y-axis, since the number of spaces must be integral. This rounding of values will lead to small misplacements of points, and straight lines may thus appear slightly distorted, although the appearance is usually acceptable. Much more precise plotting of points can be achieved using high resolution graphics (see Section 5.6).

It is useful to reserve a margin on the left-hand side of the screen in which to print numbers to indicate scale, or axis labels; 5 spaces will usually be adequate. This is the margin for showing information about the y-axis (in the conventional x–y representation); that for the x-axis can be printed below the plot.

5.2   Histogram

A simple histogram is easy to produce. Algorithm 5.1, *PlotHistogram*, assumes that a frequency table has already been formed from the data. The type of output is illustrated in Figure 5.1. The left endpoint of each interval is printed. Large frequencies will produce too many asterisks for a row and the extra asterisks for an interval affected in this way will be printed on the next line; this spoils the histogram. We suggest two alternative approaches for this case: (a) scale all the frequencies so that the maximum will fit in the space available, or (b) test whether each frequency exceeds a number which is a few units less than the maximum space available; if it does, print '+ n' at the end of the line of asterisks, n being the number of asterisks that there is not room to print.

```
procedure PlotHistogram(Frequency : FreqList;
                        LowerLimit,ClassWidth : real;
                        NoOfClasses,Margin,Dp : integer);

(* plots a histogram of data which have been classified into *)
(* a frequency table, using the values of LowerLimit,        *)
(* ClassWidth and NoOfClasses stated. The value of           *)
(* LeftEndPoint of each class is printed on the left of the  *)
(* diagram in a field of width Margin with Dp decimal places *)

var i,j : integer; (* loop counters *)
    LeftEndPoint : real; (* of class under consideration *)

begin
    for i := 1 to NoOfClasses do
    begin
        writeln(' ':Margin,' I');
        LeftEndPoint := LowerLimit + (i-1)*ClassWidth;
        write(LeftEndPoint:Margin:Dp,'-I ');
        if Frequency[i] > 0
            then for j := 1 to Frequency[i] do write(' *');
        writeln
    end; (* of printing of one line *)
    writeln(' ':Margin,' I')
end; (* of PlotHistogram *)
```

Algorithm 5.1 *PlotHistogram*

Figure 5.1

5.3 Stem and leaf diagram

Tukey (1977) has proposed a form of diagram which represents the distribution of a set of numbers in a similar way to a histogram, yet also retains most, if not all, of the information on individual values. We may think of it as a concise way of writing down a set of numbers. As an example, a stem-and-leaf diagram for the (ordered) numbers 21, 23, 25, 32, 32, 34, 35, 36, 38, 39, 40, 43, 44, 45, 51, 55, 67 is:

```
2 | 135
3 | 2245689
4 | 0345
5 | 15
6 | 7
```

```
procedure DrawStemLeaf(x : DataList; n : Units;
                       LowerLimit,ClassWidth,Margin:integer);

(* draws stem and leaf diagram with stated LowerLimit and *)
(* ClassWidth from data array x which contains n integers *)
(* with no more than three digits,in ascending order.    *)
(* Margin gives the size of the field in which the stem   *)
(* value is printed.                                      *)

var i : integer; (* label of data items *)
    xValue : integer; (* data value under consideration *)
    Leaf : integer; (* final digit of xValue *)
    LeftEndPoint : integer; (* of current class interval *)
    Maximum : integer; (* of data values *)
    RightEndPoint : integer; (* of current class interval *)
    StemLabel : integer; (* leading digit[s] of stem *)

begin
    LeftEndPoint := LowerLimit;
    xValue := x[1];
    i := 1;
    if xValue < LeftEndPoint
        then LeftEndPoint := xValue;
    RightEndPoint := LeftEndPoint + ClassWidth;
    Maximum := x[n];
    while LeftEndPoint < Maximum do
    begin
        writeln;
        StemLabel := LeftEndPoint div 10;
        write(StemLabel:Margin,' * ');
        while xValue < RightEndPoint do
        begin
            Leaf := xValue - 10*StemLabel;
            write(Leaf:1);
            i := i + 1;
            if i <= n then xValue := x[i];
        end; (* of loop adding leaves to stem *)
        LeftEndPoint := RightEndPoint;
        RightEndPoint := RightEndPoint + ClassWidth;
    end; (* of loop printing stem and leaves *)
end;
```

Algorithm 5.2 *DrawStemLeaf*

Thus 21 is represented on the first row, or **stem** as Tukey calls it, by the stem label '20', which we replace by '2', together with '1' which is a **leaf** on the stem. The other two leaves, 3 and 5, on this stem correspond to the numbers 23 and 25. The next stem has the label '30', represented by the '3' on the left, and the seven leaves on this stem correspond to the seven numbers of the set of data which are in the thirties. The idea is discussed at length, and elaborated, by Tukey (1977).

For input to Algorithm 5.2, *DrawStemLeaf*, the data are assumed to be in ascending order, having been sorted; each observation is assumed to be an integer of two or three digits. The diagram divides the data into classes. It is necessary to input the desired lower limit and class-width; in the above example the lower limit was 20 and the class-width 10. A value of 10 is particularly easy to interpret but, depending on the data, Tukey also recommends 2, 5, 20, 50 as possible class-widths. The algorithm can be used with a wide range of class-width values, and the reader may like to experiment with this on his own machine.

Data are most easily interpreted when they come from a symmetrical distribution. One use of a stem and leaf diagram is to check whether a distribution is symmetrical; if it is not, the statistician considers transforming the original data in order to obtain a more symmetrical distribution. Also, Algorithm 5.2 is rather restrictive in the type of data it will accept: it requires integers of two or three digits, and data may therefore need to be transformed into a form acceptable for this algorithm.

5.4   Box and whisker plot

Tukey (1977) proposed another form of diagram, the box and whisker plot, which may be regarded as a diagrammatic presentation of the six-number summary (see Section 4.3). (It would strictly be more accurate to say five-number summary, since the sample size is not shown unless incorporated as a separate line of output.)

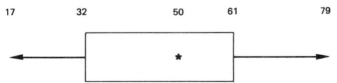

Figure 5.2

An example is shown in Figure 5.2. Basically it is a representation of the five numbers on a line, but the portion between the first quartile (32) and the third quartile (61), which contains half the population, is thickened and represented as a box. Whiskers reach out from the left and right sides of the box to the extremes (17 and 79). The median (50) is denoted by an asterisk within the box. We give an algorithm *DrawBoxPlot*, 5.3.

```
procedure DrawBoxPlot(Q : QuartileArray;
                      LowerLimit,UpperLimit : real;
                      Dp,Margin,WindowWidth : integer);

(* draws a box and whisker plot from values in the array Q.   *)
(* Q[0]=minimum value,Q[1],Q[2],Q[3] are the 1st,2nd,3rd      *)
(* quartiles, Q[4] = maximum value. Information for the       *)
(* layout of the plot is given by lower and upper end points  *)
```

```
(* of variable values, margin at left of diagram, window-   *)
(* width for diagram and number of dp in printed values.     *)
(* Procedure Line is used.                                    *)

var i : integer; (* index *)
    p : array[0..4] of integer; (* scaled quartile positions *)
    ScaleFactor : real;


procedure Line(n1,n2 : integer; Symbol : char);

(* plots a line of length n2-n1 consisting of the declared *)
(* symbol.                                                  *)

var i : integer; (* loop counter *)

begin
    if n2 >= n1
        then for i := n1 to n2 do
            write(Symbol)
end; (* of line *)


begin
    ScaleFactor := WindowWidth/(UpperLimit-LowerLimit);
    for i:= 0 to 4 do
      p[i]:=trunc(ScaleFactor*(Q[i]-LowerLimit))+Margin;
    writeln;writeln;write(' ');
    write(Q[0]:p[0]:Dp,Q[1]:p[1]-p[0]:Dp,Q[2]:p[2]-p[1]:Dp);
    writeln(Q[3]:p[3]-p[2]:Dp,Q[4]:p[4]-p[3]:Dp);
    writeln;
    Line(1,p[1]-1,' ');
    write('I');
    Line(p[1]+1,p[3]-1,'-');
    writeln('I');
    Line(1,p[1]-1,' ');
    write('I');
    Line(p[1]+1,p[3]-1,' ');
    writeln('I');
    write('(':p[0]);
    Line(p[0]+1,p[1]-1,'-');
    write('I');
    Line(p[1]+1,p[2]-1,' ');
    write('*');
    Line(p[2]+1,p[3]-1,' ');
    write('I');
    Line(p[3]+1,p[4]-1,'-');
    writeln(')');
    Line(1,p[1]-1,' ');
    write('I');
    Line(p[1]+1,p[3]-1,' ');
    writeln('I');
    Line(1,p[1]-1,' ');
    write('I');
    Line(p[1]+1,p[3]-1,'-');
    writeln('I');
end;
```

Algorithm 5.3 *DrawBoxPlot*

To scale the plot, we must input the window width (given as the number of character spaces), and the upper and lower limits of the data. These last two values may be set equal to the maximum and minimum values of the data if the plot is to occupy the whole window width. Limits distinct from the extreme data values are useful if box and whisker plots for a number of populations are to be compared; the same limits may be used in plotting each population. The number of decimal places (*Dp*) in the rounded values must also be input; unless they are rounded, the values in the six-number summary are likely to run into each other when output.

Algorithm 5.3 begins by finding the scale factor. Then the positions $p[i]$ of the five numbers are found, measured as distances from the lower limit. A number of statements can be written more concisely if we call the minimum value in the data $Q[0]$ and the maximum in the data $Q[4]$; all five summary numbers will then be of the form $Q[i]$, $i = 0, 1, 2, 3, 4$.

5.5   Scatter diagram

Scatter diagrams should always be used when investigating the relation between two variables. We illustrate in Figure 5.3 the type of output that is obtained from Algorithm 5.4, *PlotScatter*.

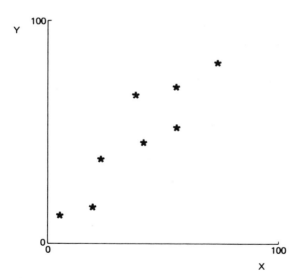

Figure 5.3

```
procedure PlotScatter(x,y : DataVector; n : Units;
                      xMin,xMax,yMin,yMax : real;
                      WindowWidth,WindowHeight : integer;
                      Margin,Dp : integer);

(* plots scattergram of n pairs of values stored in arrays x *)
(* and y; xMin,xMax and yMin,yMax give the range of x and y  *)
```

```
(* values that are plotted; the diagram appears in a window   *)
(* of stated width and height (in units of print positions);  *)
(* to the left is a margin of given width; numerical values   *)
(* on the axes are printed correct to a stated number of      *)
(* decimal places(Dp); multiple points on the  diagram are    *)
(* represented by the number of points except that a  @  is   *)
(* used when the number exceeds 9.                            *)

var FreqGrid : FreqTable; (* frequency table of scaled data *)
                         (* values.                         *)
    Frequency : integer; (* entry in frequency table *)
    i : integer; (* loop counter *)
    xClass,yClass : integer; (* plotted class intervals *)
    xRange,yRange : real; (* extents of x and y plots *)
    xScaleFactor,yScaleFactor : real; (* scale factors used *)
                                     (* to plot diagram *)

begin
    xRange := xMax - xMin;
    yRange := yMax - yMin;
    for xClass := 1 to WindowWidth do
        for yClass := 1 to WindowHeight do
            FreqGrid[xClass,yClass] := 0;
    xScaleFactor := (WindowWidth - 1)/xRange;
    yScaleFactor := (WindowHeight - 1)/yRange;
    for i := 1 to n do
    begin
        xClass := trunc((x[i] - xMin)*xScaleFactor) + 1;
        yClass := trunc((y[i] - yMin)*yScaleFactor) + 1;
        FreqGrid[xClass,yClass] := FreqGrid[xClass,yClass] + 1
    end;
    for yClass := WindowHeight downto 1 do
    begin
        if yClass = WindowHeight
            then write(yMax:Margin-1:Dp,'-')
            else write(' ':Margin);
        write('I':1);
        for xClass := 1 to WindowWidth do
        begin (* plot points *)
            Frequency := FreqGrid[xClass,yClass];
            if Frequency < 1
                then write(' ':1)
                else if Frequency > 9
                    then write('@':1)
                    else write(Frequency:1)
        end;
        writeln
    end; (* of loop indexed by yClass *)
    write(yMin:Margin-1:Dp,'-+');
    for i := Margin+2 to WindowWidth+Margin do
        write('-');
    writeln;
    writeln(xMin:Margin+Dp+1:Dp,xMax:WindowWidth+1:Dp);
    writeln
end; (* of PlotScatter *)
```

Algorithm 5.4 *PlotScatter*

The method adopted is, first, to classify the data into a two-way frequency table, and then to print a mark on the plot corresponding to cells which have a non-zero frequency. When the frequency is unity an asterisk is printed, and when it is greater than unity the actual value of the frequency is printed.

The method used in the first part of the algorithm is a simple extension to two variables of Algorithm 4.4, *FormFreqTable*. The second part of Algorithm 5.4 prints the axes, the numerical values at the extremes of the axes, and the points.

5.6   High resolution graphics

The graphical methods we have discussed rely on a grid of character print positions; this might, for example, be a grid of 24 rows × 80 columns. But each character is formed by setting dots 'on' or 'off' within a matrix of, say, 7 × 8 dots. Graphical package programs exist for many computers, which may be linked in with Pascal to give control over individual printing of the dots we have referred to. For our example values, this would allow control of 168 × 640 print positions. This enables the programmer to produce much finer pictures and to achieve much closer approximations to continuous straight lines. In conjunction with this high resolution, there are inbuilt functions which will draw a straight line between two points whose coordinates are given. This opens up enormous possibilities in representing data pictorially. Conventional histograms consisting of rectangles are easy to construct, and pie charts also become possible; even on moderately priced machines the segments of pie charts can be exhibited in contrasting colours.

5.7   Test data for Algorithms

Algorithm 5.1, *PlotHistogram*. Input: $NoOfClasses = 5$, $Frequency = (2, 4, 6, 12, 6)$, $LowerLimit = 5.14$, $ClassWidth = 2.5$, $Dp = 1$, $Margin = 10$.
Algorithm 5.2, *DrawStemLeaf*. Input: $n (= 30)$ numbers of the form $x[i] = a + b * i$, where $a = 1$, $b = 3$ and $i = 1$ to n. $LowerLimit = 0$, $ClassWidth = 10$.
Algorithm 5.3, *DrawBoxPlot*. Input: $Quartile = (20, 40, 51.5, 73, 120)$, $LowerLimit = 10$, $UpperLimit = 150$, $Dp = 1$, $Margin = 4$, $WindowWidth = 40$.
Algorithm 5.4, *PlotScatter*. Input: Pairs $(x, y) = (2, 2), (7, 2), (7, 7), (12, 2), (12, 7), (12, 12)$. $n = 6$, $xMin = 1$, $xMax = 15$, $yMin = 2$, $yMax = 20$, $WindowWidth = 30$, $WindowHeight = 20$, $Margin = 4$, $Dp = 0$. Output: triangle of points.

5.8   Exercises

1   At the beginning of Algorithm 5.2, *DrawStemLeaf*, introduce a procedure which:
   i) checks that the data in x are in ascending order, and sets a boolean variable *DataAscending* to 'true' if that is so, and to 'false' otherwise;
   ii) checks that *LowerLimit* is less than the minimum of the data, i.e. $x[1]$, and sets a boolean variable *DataInRange* to 'true' if that is so, but to 'false' otherwise. Let *DrawStemLeaf* continue only if both *DataAscending* and *DataInRange* are true. Otherwise return to the calling program.

2† Write a program which:
i) inputs *n* data, each of which is an integer not exceeding three digits, into an array *x*;
ii) sorts the data and puts them in *x* in ascending order of size;
iii) finds the maximum and minimum of the data;
iv) uses *DrawStemLeaf* (Algorithm 5.2) with *LowerLimit* = 0 and
ClassWidth = *trunc* (($Maximum - Minimum$)/12).

3 Write an algorithm to compare *two* sets of data using stem-and-leaf plots. A convenient display is to have a central 'stem' column with the 'leaves' for one set of data to the right (as in Algorithm 5.2) and the leaves for the other set of data to the left of the central column.

4† Write a program which:
i) inputs *n* data into an array *x*;
ii) calculates the six-number summary (see Section 4.3);
iii) prints a box and whisker plot (see Section 5.4);
iv) allows a further set of data to be entered, and an additional box and whisker plot to be printed, which may be compared with the previous plot.

5 Use *PlotScatter* (Algorithm 5.4), to plot a *time-series*, i.e. to show how the record Y varies when X represents equally-spaced time intervals, as in the following data on the quarterly production of a certain animal feeding stuff (in tonnes).

Year/Quarter	Production	Year/Quarter	Production
1977/1	32.5	1979/1	35.4
2	42.2	2	48.0
3	35.1	3	42.6
4	41.0	4	49.8
1978/1	34.7	1980/1	38.5
2	45.5	2	52.3
3	37.6	3	47.3
4	44.2	4	54.3

6 Run Algorithm 5.2. *DrawStemLeaf*, using as input data *either* (a) the atomic masses of the elements taken from a table in a chemistry book (take the stem in units of 10) *or* (b) random numbers between 0 and 100 generated using the *Random* function, (Algorithm 7.2), except that the value should be multiplied by 100.

7 Use appropriate algorithms to construct a scatter diagram for the following data, which give the mathematics (X) and physics (Y) marks of eleven students A–L in the same class:

	A	B	C	D	E	F	G	H	J	K	L
X:	41	37	38	39	49	47	42	34	36	48	29
Y:	36	20	31	24	37	35	42	26	27	29	23

8† Repeat question 7 for suitable pairs of data from Appendix B, e.g. (a) H and W, (b) R and C, separately for the two sexes.

6 Computation of variance and correlation coefficients

6.1 Variance by methods based on standard formulae

Texts on statistics usually give two alternative and equivalent formulae for calculating the estimated population variance, s^2, based on a sample of n observations:

$$s^2 = \sum_{i=1}^{n} (x_i - \bar{x})^2 / (n-1); \tag{1}$$

$$s^2 = \left(\sum_{i=1}^{n} x_i^2 - (\sum x_i)^2 / n \right) / (n-1). \tag{2}$$

Note that we are using a divisor $(n-1)$. This is appropriate when estimating the variance of a population, given a sample of n observations from the population (*ABC*, Section 10.5), and is what we usually require. If the n observations do actually constitute a complete population, the variance is given by similar formulae in which n replaces $(n-1)$. Formula (1) is the direct definition of variance, while in (2) the sum of squares $\Sigma(x_i - \bar{x})^2$ has been expanded and rearranged in a form that is often more convenient for desk calculation. The term $(\sum x_i)^2 / n$ is called a **correction term** and the numerator $\sum x_i^2 - (\sum x_i)^2 / n$ is called a **corrected sum of squares**. Taking the square root of the variance gives the standard deviation.

```
procedure FindMeanVar1(x : DataVector;   n : Units;
                       var Mean,Variance : real );

(* calculates variance,with divisor n-1, and mean of n     *)
(* observations in an array x using the two-pass deviation  *)
(* method  *)

var   Dev : real; (* deviation of observation from the meam *)
      i : integer; (* counter *)
      Sum : real; (* temporary store for summing values *)
begin
    Sum := 0.0;
    for i := 1 to n do
    Sum := Sum + x[i];
    Mean := Sum/n;
    Sum := 0.0;

    for i := 1 to n do
    begin
        Dev := x[i] - Mean;
        Sum := Sum + Dev*Dev
    end; (* of summing squares of deviations *)
    Variance := Sum/(n - 1)

end; (* of FindMeanVar1 *)
```

Algorithm 6.1 *FindMeanVar1*

Algorithm 6.1, *FindMeanVar1*, is constructed by the method of formula (1). This algorithm is straightforward and easy to understand. It might be criticised on grounds of efficiency, as it requires two passes through the data: one to calculate the mean and a second to calculate the squared deviations from the mean, which then allow us to calculate the variance.

Formula (2) looks as if it might be more efficient since the calculation requires only a single pass through the data. Unfortunately it has a serious disadvantage: it can give inaccurate results. The subtraction of the correction term from the uncorrected sum of squares $\sum x_i^2$ may lead to a substantial loss of significant figures because it takes the difference of two numbers, both of which may be large. We illustrate the nature of the inaccuracy by an example.

Suppose we wish to estimate the variance from a sample consisting of the numbers 1000, 1003, 1006, 1007, 1009, using a computer which expresses floating point numbers correct to 6 decimal digits. In Table 6.1 we follow through the calculation of the variance using formula (2), assuming (a) that the numbers are truncated, or (b) that they are rounded, in order to fit into the space available for each number. We give also the exact calculation; and we see that the result is disconcerting. The errors in the results (7.5 or 15, instead of the correct value of 12.5) are substantial; and other sets of numbers could be chosen that would give even more dramatic errors. An operator using a desk calculator would immediately 'code' the data by subtracting 1000 from each observation before calculating a variance, and this would avoid the error. But one cannot rely on data being first scrutinised like this when a computer is used. We must conclude that the method based on formula (2) can be unreliable and so should not be used.

Table 6.1

	Exact	Truncated	Rounded
x_1^2	1000000	100000 E1	100000 E1
x_2^2	1006009	100600 E1	100601 E1
x_3^2	1012036	101203 E1	101204 E1
x_4^2	1014049	101404 E1	101405 E1
x_5^2	1018081	101808 E1	101808 E1
$A = \sum x^2$	5050175	505015 E1	505018 E1
$(\sum x)^2$	25250625	252506 E2	252506 E2
$B = (\sum x)^2/5$	5050125	505012 E1	505012 E1
$A - B$	50	3 E1 = 30	6 E1 = 60
Variance, $(A - B)/4$	12.5	7.5	15

We recommend the two-pass deviation method as the best method for general use. An improvement in accuracy, in situations where the standard deviation is small compared with the mean, may be obtained by making a correction to the sum of squares of deviations. The formula is:

$$s^2 = \left(\sum_{i=1}^{n} (x_i - \bar{x})^2 - \left(\sum_{i=1}^{n} (x_i - \bar{x}) \right)^2 / n \right) / (n-1).$$

The correction is zero in exact computation, but in normal computation is a good approximation to the error in the first term.

There may be situations in which it is considered important to calculate the variance in a single pass. We give a method of doing this in Section 6.2. Chan *et al* (1983) discuss the merits of alternative methods of calculating the variance.

6.2   An updating procedure for calculating mean and variance

This method updates the value of the mean and the sum of squares of deviations as each observation is introduced into the calculation. However, it is a less obvious and straightforward method than that of Algorithm 6.1, based on the direct definition of variance. It is also a less accurate computer method.

Write m_i for the mean of the first i observations ($= \sum_{r=1}^{i} x_r / i$), and s_i for the sum of squares of deviations of the first i observations about their mean ($= \sum_{r=1}^{i} (x_r - m_i)^2$). We shall prove, in Section 6.3, the recurrence relations

$$m_i = [(i-1)m_{i-1} + x_i]/i;$$
$$s_i = s_{i-1} + (i-1)(x_i - m_{i-1})^2/i.$$

Algorithm 6.2, *FindMeanVar2*, calculates the mean and variance of a set of data, based on these recurrence relations.

```
procedure FindMeanVar2(x : DataVector; n : Units;
                       var Mean,Variance : real );

(* calculates variance,with divisor  n - 1, and mean of n *)
(* observations in an array x using the one-pass update   *)
(* method.                                                 *)

var     Dev : real; (* deviation of observation from mean *)
        i : integer; (* counter *)
        Sum : real; (* temporary store for summing values *)

begin
    Sum := 0.0;
    Mean := 0.0;
    for i := 1 to n do
    begin
        Dev := x[i] - Mean;
        Mean := ((i - 1)*Mean+x[i])/i;
        Sum := Sum + Dev*Dev*(i - 1)/i
    end; (* of loop *)
    Variance := Sum/(n - 1)
end; (* of FindMeanVar2 *)
```

Algorithm 6.2 *FindMeanVar2*

6.3   Proof of recurrence relations for mean and variance

We require to prove the two relations given in Section 6.2. The proof of that for m_i is straightforward. If we multiply both sides of the equation by i, we have

$$im_i = (i-1)m_{i-1} + x_i.$$

Each side is now the sum of the i observations, since

$$im_i = \sum_{r=1}^{i} x_r \quad \text{and} \quad (i-1)m_{i-1} = \sum_{r=1}^{i-1} x_r.$$

Before proving the relation for s_i, note that

$$
\begin{aligned}
i(m_i - m_{i-1}) &= im_i - im_{i-1} \\
&= (i-1)m_{i-1} + x_i - im_{i-1} \quad \text{(from the result for means)} \\
&= x_i - m_{i-1}.
\end{aligned}
$$

Now

$$
\begin{aligned}
s_{i-1} &= \sum_{r=1}^{i-1} (x_r - m_{i-1})^2 = \sum_{r=1}^{i} (x_r - m_{i-1})^2 - (x_i - m_{i-1})^2 \\
&= \sum_{r=1}^{i} [(x_r - m_i) + (m_i - m_{i-1})]^2 - (x_i - m_{i-1})^2 \\
&= \sum_{r=1}^{i} (x_r - m_i)^2 + \sum_{r=1}^{i} (m_i - m_{i-1})^2 - (x_i - m_{i-1})^2.
\end{aligned}
$$

Here the cross-product term disappears since

$$
\begin{aligned}
\sum_{r=1}^{i} (x_r - m_i)(m_i - m_{i-1}) &= (m_i - m_{i-1}) \sum_{r=1}^{i} (x_r - m_i) \\
&= (m_i - m_{i-1}) \times 0 = 0.
\end{aligned}
$$

Thus

$$
\begin{aligned}
s_{i-1} &= s_i + \sum_{r=1}^{i} (x_i - m_{i-1})^2/i^2 - (x_i - m_{i-1})^2 \\
&= s_i + (x_i - m_{i-1})^2/i - (x_i - m_{i-1})^2 \\
&= s_i - (i-1)(x_i - m_{i-1})^2/i
\end{aligned}
$$

which leads at once to the result in Section 6.2.

6.4   Calculation of sums of squares and products of two variates

To calculate the product-moment correlation coefficient of two variates X and Y, we need the corrected sums of squares, $SSx = \sum (x_i - \bar{x})^2$ and $SSy = \sum (y_i - \bar{y})^2$, and the corrected sum of products $SPxy = \sum (x_i - \bar{x})(y_i - \bar{y})$. Algorithms 6.1 and 6.2 include the calculation of a corrected sum of squares; Algorithm 6.1 extends in an obvious way to give Algorithm 6.3, *FindSSandSP*, which also includes the calculation of the corrected sum of products.

```
procedure FindSSandSP(x,y : DataVector; n : Units;
                      var xMean,yMean,SSx,SSy,SPxy : real);

(* finds the corrected sums of squares SSx,SSy and corrected *)
(* sum of products SPxy from n observations in each of the   *)
(* arrays x and y.                                           *)

var    xDev,yDev : real; (* deviations of x and y values from *)
                         (* their means *)
       i : integer; (*  counter *)
       xSum,ySum,xySum : real; (* temporary accumulators *)

begin
    xSum := 0.0;
    ySum := 0.0;
    for i := 1 to n do
    begin
        xSum := xSum + x[i];
        ySum := ySum + y[i]
    end;
    xMean := xSum/n;
    yMean := ySum/n;

    xSum := 0.0;
    ySum := 0.0;
    xySum := 0.0;
    for i := 1 to n do
    begin
        xDev := x[i] - xMean;
        yDev := y[i] - yMean;
        xSum := xSum + xDev*xDev;
        ySum := ySum + yDev*yDev;
        xySum := xySum + xDev*yDev
    end;
    SSx := xSum;
    SSy := ySum;
    SPxy := xySum

end; (* of FindSSandSP *)
```

Algorithm 6.3 *FindSSandSP*

A general algorithm to calculate corrected sums of squares and products for many variates is given on page 156 (Algorithm A.3).

To extend the updating method of Section 6.2, we need to find a recurrence relation for the sum of products. By analogy with the definition and relation for s_i, we define the sum of products of i pairs of observations by

$$p_i = \sum_{r=1}^{i} (x_r - m_{1,i})(y_r - m_{2,i}),$$

and it can be shown that

$$p_i = p_{i-1} + (i-1)(x_i - m_{1,i-1})(y_i - m_{2,i-1})/i.$$

6.5   Input of bivariate data

Bivariate data are pairs of measurements, each pair being recorded on a single unit of a population or sample. Thus we might measure heights and weights of a sample of men; a man is a unit and the two variates are height and weight. For small sets of data, say less than 20 units (pairs), the data might be input using Algorithm 6.4, *InputBivariateData*, reading over the data on entry as a check.

```
procedure InputBivariateData(var x,y : DataVector;
                             var n : Units);

(* input of a number n and of data pairs stored in arrays x *)
(* and y respectively, each of n elements                   *)

var i : integer; (* label of data pair *)

begin
    write('State number of data pairs ');
    readln(n);
    writeln('Input pairs of observations in free format');
    writeln('Check data on each line before pressing RETURN');
    for i := 1 to n do
        read(x[i],y[i]);
    readln
end; (* of InputBivariateData *)
```

Algorithm 6.4 *InputBivariateData*

The input of larger sets of data should incorporate fuller checks. This is discussed in Chapter 13.

6.6   Calculation of product-moment correlation coefficient

To obtain the correlation coefficient *r* is a one-line calculation when we have the results of an algorithm (such as 6.3) for finding sums of squares and products (*ABC*, Section 20.3);

$$r: = SPxy/sqrt(SSx * SSy);$$

It is always wise, before calculating a correlation coefficient, to plot a scatter diagram to check whether the points representing the data pairs (x_i, y_i) are approximately linearly placed, or whether there is any systematic curved trend. We recommend that the reader should always do this.

6.7   Calculation of Spearman's rank correlation coefficient

This calculation is made from *n* pairs of ranks. For example, a teacher might rank his class of pupils according to (i) their mathematical ability and (ii) their musical ability.

He might assign ranks from his general knowledge of the pupils, or by putting in order the marks from examinations in mathematics and music. The calculation is exactly the same as for the product-moment correlation, with the ranks of one variate (e.g. mathematical ability) put in array *x* and the ranks of the other variate (e.g. musical ability) put in array *y* (*ABC*, Section 20.9).

6.8   The menu method of assembling algorithms

Analysing data is not a straightforward procedure like solving a quadratic equation. Data are investigated in various ways; the tools used are tables, graphs, summary statistics and statistical tests. The outcome of one procedure may influence which other procedure is used next. The data may be modified, as the analysis proceeds, by rejecting some items or by the use of a mathematical transformation. Thus a computer program in which there is a fixed sequence of algorithms is not likely to be convenient for a complete statistical analysis. We here suggest a form of program that allows an interactive type of analysis.

The essence of the method is that the program offers a list of procedures, commonly called a **menu**, of which one is chosen, carried out, and followed by a return to the menu. Each time the menu is consulted another procedure may be chosen or the analysis stopped. In the simple example we present in Program 6.5, *MeanVarPackage*, two procedures are offered: (1) input of data; (2) calculation of mean and variance. The program might be used to calculate the mean and variance of each of a number of data sets. The program should be self-explanatory.

If a large set of data is to be worked over many times it will be worth putting the data into a data file on tape or disk. Procedures for setting up files are discussed in Chapter 13.

```
program MeanVarPackage(input,output);

(* offers choice of two procedures, InputData and   *)
(* FindMeanVar1, to illustrate the use of a menu.    *)

const MaxSampleSize = 50;
      NoOfProcedures = 2;

type  CharacterSet = set of char;
      Units = 1..MaxSampleSize;
      DataVector = array[Units] of real;

var   Choice : integer; (* index number of procedure chosen *)
      Mean : real;
      n : Units; (* sample size *)
      ReplySet : CharacterSet; (* set of possible replies *)
      Response : char; (* response to question *)
      Variance : real;
      WantMenu : boolean;
      x : DataVector; (* data array *)


procedure PrintMenu(var Choice : integer);

var   Response : char;
```

```
begin
    repeat
        write('Type number of procedure required and RETURN ');
        readln(Choice);
    until Choice in [1..NoOfProcedures];
end; (* of PrintMenu *)


procedure InputData(var x : DataVector; var n : Units);

        (***     Algorithm 4.1    ***)



procedure FindMeanVar1(x : DataVector; n : Units;
                       var Mean,Variance : real);

        (***     Algorithm 6.1    ***)



begin  (* main programme segment *)
    ReplySet := ['y','Y','n','N'];
    repeat
        repeat
            writeln;writeln('Do you want the menu ?');
            write('Type y for yes or n for no and RETURN  ');
            readln(Response);
        until Response in ReplySet;
        WantMenu :=  Response in ['y','Y'];
        if WantMenu
        then begin
            PrintMenu(Choice);
            case Choice of
                1 : InputData(x,n);
                2 : begin
                        FindMeanVar1(x,n,Mean,Variance);
                        writeln('Number of observations = ',n);
                        writeln('Mean             = ',Mean);
                        writeln('Variance         = ',Variance);
                    end
            end;
        end; (* of choice of procedure *)
    until not WantMenu;
end.
```

Program 6.5 *MeanVarPackage*

6.9   Test data for Algorithms

Algorithm 6.1, *FindMeanVar1*. Input: $n = 5$, $x = (1, 2, 3, 5, 9)$. Output: *Mean* $= 4$, *Variance* $= 10$.

Algorithm 6.2, *FindMeanVar2*. The same as for Algorithm 6.1 above.

Algorithm 6.3, *FindSSandSP*. Input: $n = 5$, Data pairs (x, y): (1, 3), (2, 7), (3, 11), (5, 19), (9, 5). Output: *xMean* $= 4$, *yMean* $= 9$, *SSx* $= 40$, *SPxy* $= 10$, *SSy* $= 160$.

6.10   Exercises

1   Using Algorithm 6.1, *FindMeanVar1*, construct a program which will print out the mean and the standard error of the mean, which equals *sqrt*(*Variance*/*n*) of as many sets of data as required. Test the program with the following data sets.
1) Number of children in family: 3, 1, 1, 2, 7.
2) Reaction time (in milliseconds): 217, 161, 198, 242, 140.
3) Age (in years): 19.5, 19.7, 20.1, 21.1, 22.6.

2   Run Algorithm 6.1, *FindMeanVar1*, with data sets of the following form:

$$1 + 10^{-k}, 1 + 2 \times 10^{-k}, 1 + 3 \times 10^{-k}, 1 + 5 \times 10^{-k}, 1 + 9 \times 10^{-k}$$

in which k takes the values 1, 2, 3, ... Thus the first three data sets are:
1) 1.1, 1.2, 1.3, 1.5, 1.9;
2) 1.01, 1.02, 1.03, 1.05, 1.09;
3) 1.001, 1.002, 1.003, 1.005, 1.009.

The exact values of mean and estimated variance are, respectively, $1 + 4 \times 10^{-k}$ and 10^{1-2k}. Note the first value of k at which inexact values are printed, and continue until ridiculous values are obtained.

3   Write a program to calculate and print the mean and estimated variance of a set of data, using formula (2) of Section 6.1 to calculate the variance. Test this program using the data sets described in question 2.

4   Add extra instructions to Algorithm 6.1, *FindMeanVar1*, so that the variance is also calculated using the correction described at the end of Section 6.1. Investigate the results with various sets of data to discover when the correction has an effect.

5   In the program of question 1, print the standard error to *Sf* significant figures, and print the mean to the same number of *decimal places* as the standard error. For most data, *Sf* = 2 will be suitable; for very accurately recorded data put *Sf* = 3.

6   In quality control, where observations are collected at fixed (equal) time intervals, the random component of variation is often estimated by

$$s^2 = \frac{1}{2(n-1)} \sum_{i=1}^{n-1} (x_{i+1} - x_i)^2$$

where x_i is the observation recorded at time i, and $i = 1, 2, \ldots, n$. Write a program to calculate s^2, and to keep it updated as fresh observations are recorded.

7   Write a program to calculate the correlation coefficient between x and y given n pairs of data (x, y), using Algorithm 6.3.
  Use the program to calculate the correlation coefficient from n pairs of data with $x = i$, $y = i(10 - i)$, and $i = 1$ to n. Print out the values of the coefficient as n takes the values 4, 5, 6, 7, 8, 9, and note the influence of the number of pairs on the value of the coefficient.

8   Write a procedure to calculate sums of squares and products as in Algorithm 6.3, *FindSSandSP*, but which uses the update method described at the end of Section 6.4.

9† *Spearman's Rank Correlation Coefficient.* There are two methods of carrying out this calculation; one is described in Section 6.7 and the other in *ABC*, Section 20.9. The methods of constructing programs to calculate the coefficients are as follows.

(A) (based on Section 6.7).
  (1) Input the number of units n, and the paired data arrays x, y, each of n elements.
  (2) Calculate the ranks of x using *FindRanks* (Algorithm 3.5) as a procedure, and put the ranks in x.
  (3) Repeat (2) on y.
  (4) Calculate SSx, SSy and $SPxy$, using Algorithm 6.3.
  (5) Print the rank correlation coefficient

    $SPxy/sqrt(SSx*SSy)$.

(B) (based on the formula $1-6\sum d_i^2/n(n^2-1)$, where d_i is the difference in the ranks of x, y for unit i; see *ABC*, Section 20.9).
  (1), (2), (3) as in (A) above.
  (4) Calculate $T = \sum d_i^2$, where $d_i = x[i] - y[i]$, the difference of the *ranks* of the two variables.
  (5) Print the rank correlation coefficient

    $1-6*T/(n*(n*n-1))$.

Write a program for each of these methods, and use it on the data in *ABC*, Section 20.9, on the rankings given to strawberry jam samples by two judges:

Sample	i	ii	iii	iv	v	vi	vii	viii	ix	x
Judge A's rank (X)	7	8	1	6	3	9	2	4	5	10
Judge B's rank (Y)	10	9	4	3	6	8	1	2	5	7

10 Write a program to calculate the mean and variance of a set of data which are given in the form of a grouped frequency table. Assume that the ends of each interval are given; calculate the mid-points $x[i]$ of the intervals, and let $f[i]$ be the frequency of observations in interval i, with i taking values 1 to n. Then the mean and variance are found from the formulae (*ABC*, Section 4.5.2):

$$\bar{x} = \sum_{i=1}^{n} f_i x_i \Big/ \sum_{i=1}^{n} f_i$$

$$\text{and} \quad s^2 = \sum_{i=1}^{n} f_i(x_i - \bar{x})^2 \Big/ \left(\sum_{i=1}^{n} f_i - 1\right).$$

The formula for s^2 is the extension of formula (1) in Section 6.1. Your program can therefore be regarded as the extension of Algorithm 6.1.

11 Extend Program 6.5, *MeanVarPackage*, by adding more procedures. Possible extensions are: (i) a procedure which replaces each observation x by a power of x, e.g. x^c where $-3 \le c \le 3$, and $c = 0$ corresponds to $\ln(x)$; (ii) a six-number summary (see Program 4.2); (iii) *FormFreqTable* and *OutputFreqTable* (Algorithms 4.4 and 4.5); (iv) *PlotHistogram* (Algorithm 5.1).

7 Simulation

7.1 Introduction

A favourite game on computers is 'lunar lander'. In one version there is a lunar landscape displayed at the bottom of a VDU screen and a spacecraft appears at the top of the screen. The aim of the player is to make a gentle landing of the spacecraft on the moon's surface. Control of the spacecraft's movement is exercised by modifying the thrust of the two engines of the spacecraft; the player is able to input this information from the keyboard. The program in the computer calculates the position at time $t + 1$ given the thrust of the engines, and the position and velocity of the spacecraft at time t. The motion of a real spacecraft is thus mimicked by the computer; we call this **simulation**. In a simulation we pretend something is happening and work through the essential stages of the process, often using a mathematical model.

Computers make simulation so easy that it is a technique widely used in, for example, industry, business, teaching and research. We give only a brief introduction to the subject in this chapter and in Chapter 9; it may be pursued in Tocher (1963), Fishman (1978), Morgan (1984) and Yakowitz (1977).

The 'lunar lander' game described above is a **deterministic** simulation. When we know the state of the spacecraft (i.e. its position and velocity and the thrust of its engines) at time t, then its state at time $t + 1$ is completely determined; no chance or random element is involved in the program. We shall now consider in detail a **stochastic** simulation, where a chance element is involved. A large part of the chapter is given over to discussing the generation of random numbers, the means by which the chance element is brought into a simulation; see Knuth (1981) and Ripley (1983) as well as references listed above.

7.2 Simulation of card collecting

Cigarette manufacturers used to give away cards in packets of cigarettes. Each card was from a set whose subject might be famous cricketers or film stars or breeds of dogs. A desire to complete the set led smokers to buy more cigarettes; other manufacturers use the same idea today. Using a computer simulation, it is easy to mimic how a collection would be built up. Probably the question of most interest is: 'What is the distribution of the number of packets that need to be bought to complete a set of cards?' This may be answered easily from the simulation. The core of the simulation is shown in a flow diagram in Figure 7.1. A program for the simulation is given in Program 7.1 (page 63). Suppose the number of cards in a complete set is *NoInSet*. The state of the collector's set at any time is recorded in an array called *Want* with *NoInSet* elements; initially each element is set to 'true' and when the *i*th card is collected *Want*[*i*] is changed to 'false'. A

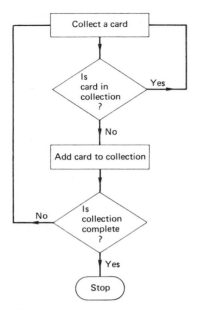

Figure 7.1

record is kept of the number of cards wanted; when the number is zero a simulation ends.

Collecting a card is represented by choosing a random integer between 1 and *NoInSet* inclusive. A function *Random* (Algorithm 7.2) generates a number between 0 and 1, and the value is intended to be a random sample from the continuous uniform distribution (*ABC*, Section 13.5) on the interval (0, 1). (The generation of random numbers by computer is discussed in Section 7.3.) Each time the function is called a new random sample is chosen from (0, 1). This is used to produce a random integer between 1 and *NoInSet* inclusive.

```
program CardCollection(input,output);

(* simulation of collecting cards from packets to make *)
(* up a set; counts number of packets bought.          *)

var CardNo,RunCount : integer;
    Frequency : array[1..100] of integer; (* number of a *)
                             (* particular card collected   *)
    i : integer; (* loop counter *)
    NoOfPackets,NoInSet,NoOfRuns,NoWanted : integer;
    Seed : integer; (* seed for random number generator *)
    Want : array[1..100] of boolean; (* indicates that a *)
                             (* particular card is required *)

function Random(var w : integer) : real;

          (***  Algorithm 7.2  ***)
```

```
begin    (* set parameters *)
    writeln; writeln;
    write('number of simulations  ');
    read(NoOfRuns);
    write('number in set           ');
    read(NoInSet);
    writeln('seed for random number generator ');
    write('type any positive integer        ');
    read(Seed);
    RunCount := 0;
    writeln; writeln;
    repeat  (* start a collection  *)
        RunCount := RunCount + 1;
        for i := 1 to NoInSet do
        begin
            Want[i] := true;
            Frequency[i] := 0;
        end;
        NoOfPackets := 0;
        NoWanted := NoInSet;
        repeat  (* obtain a card  *)
            CardNo := 1 + trunc(NoInSet*Random(Seed));
            Frequency[CardNo] := Frequency[CardNo] + 1;
            NoOfPackets := NoOfPackets + 1;
            if Want[CardNo] = true
            then begin
                Want[CardNo] := false;
                NoWanted := NoWanted - 1
            end;
        until NoWanted = 0;
        write('simulation ',RunCount:4);
        writeln('    no of packets  ',NoOfPackets:4);
    until RunCount = NoOfRuns;
end.
```

Program 7.1 *CardCollection*

The total number of cards inspected to complete the set (which equals the number of packets purchased) is counted and printed. The number of cards of each sort is counted in the array *Frequency*; this information may be printed if required. Ten runs of the program with 30 cards in a set gave the number of packets as:

109, 99, 131, 114, 138, 162, 149, 112, 147, 113.

These numbers illustrate a common characteristic of stochastic simulations: the results are often very variable.

7.3   Generation of pseudo-random numbers

Standard Pascal does not include a random number generator, though the reader may find that a function is available on his machine to generate the random numbers he requires for simulations. But random number generators are not perfect so it is wise to understand the methods that are used and to appreciate their limitations.

Before computers became widely available, statisticians used tables of random digits to choose random samples from populations or distributions. A line of such a table might take the form:

17 53 77 58 71   71 41 61 50 72   12 41 94 96 26   44 95 27 36 99

One might consider storing a table of random digits in a computer but currently no-one does so. To be speedily available they would have to be stored in the central part of the computer, and programmers consider that to be a waste of precious storage space. If, alternatively, the numbers were put in an ancillary storage device such as a disk or cassette, retrieval would be too slow. What is desired is that the computer shall produce lots of random numbers on demand, using a simple program that takes up little storage space. The successive digits in the decimal expansion of π come to mind as digits that have no obvious pattern and that we might therefore consider 'random', although their generation would take too long for it to be a convenient method of obtaining random digits, even if it satisfied the other necessary criteria. A moment's thought makes us ask whether we are demanding the impossible. We want the numbers to be generated by a short program and hence by a completely deterministic process. How can they possibly be *random* numbers? We have to admit that they cannot be random numbers in any strict sense (although investigation has shown that it is difficult to say exactly what a random number is) and that all the current random number generators have faults. The aim is to produce numbers that have the properties we expect random numbers to have, if they are to be satisfactory in the application we have in mind. A more correct term for the numbers produced is **pseudo-random** numbers, though for brevity we shall refer to them as random numbers.

For non-computer use, statisticians found random *digits* the most convenient form of random number. The numbers were random samples from a discrete uniform distribution (*ABC*, Section 7.4) taking values 0, 1, 2, . . . , 9 each with probability 0.1. The digits were combined as required. The most convenient random numbers for computer use, however, are random samples from the continuous uniform distribution (*ABC*, Section 13.5) on the interval (0, 1). We may represent this distribution by U (0, 1), and a typical value by u. The values are of course truncated since the computer can hold numbers only to a fixed number of places. Examples of pseudo-random numbers generated by a computer are:

.131 137 465,   .809 248 730,   .846 447 204,   .841 536 558,   .591 965 711,
.268 001 130.

Random digits, if required, are generated from a random number u by the instruction:

   RandomDigit: $= trunc(10 * u)$.

We shall see later (Chapter 9) that random samples from U (0, 1) are the basis for obtaining random samples from any statistical distribution.

To sum up our requirements for the pseudo-random numbers: we wish to generate efficiently (both in speed and storage) numbers that we may regard as independent (an alternative but more precise word than 'random') samples from U(0, 1).

7.3.1   The mid-square method

One of the first approaches to this problem was that of John von Neumann, a mathematician who made many early contributions to the theory of computing. A p-digit number w_0 is squared. The central p digits of w_0^2 are taken as w_1, which in turn is squared and the resulting central digits taken as w_2, etc. With $p = 4$, we might begin with $w_0 = 7137$ to give $w_0^2 = 50\,936\,769$, yielding $w_1 = 9367$. The sequence obtained, after dividing each number by 10^4 to obtain values in the interval (0, 1), is

$$.7137, .9367, .7406, .8488, .0461, .1215, .5156, \ldots$$

There are 10 000 possible values of w (using $p = 4$). Hence at some stage a number will repeat, and from then on the whole sequence will repeat. The maximum possible length of a cycle would be 10 000 but cycles are in practice much shorter. For example, with $w_0 = 1357$ we obtain the following sequence of 4-digit numbers:

$$1357, 8414, 7953, 2502, 2600, 7600, 7600, 7600, \ldots.$$

The sequence settles into a cycle of period 1. Another sequence is

$$9999, 9800, 0400, 1600, 5600, 3600, 9600, 1600, \ldots$$

which yields a cycle of period 4. These features are not satisfactory. Moreover it can be shown (Tocher, 1963) that the 10^p possible values are not equally likely; therefore the method fails in one of the basic requirements of a random sequence. But it does suggest a way of proceeding. We seek to generate a sequence $w_i (i = 0, 1, 2, \ldots)$ where each term is generated from the previous one by a relation $w_{i+1} = f(w_i)$ such that the cycle length is very long.

7.4   The congruence method

Another mathematician, Lehmer, suggested using the theory of numbers to devise the required sequences; in particular, using the theory of congruences. The generating relation is

$$w_{i+1} = a_1 w_i + a_2 \quad (\text{mod } k)$$

where all the quantities are integers. The right-hand side $(a_1 w_i + a_2)$ is evaluated, and working modulo k means that the expression is replaced by the remainder after dividing by k. Let us use simple numbers $(a_1 = 21, a_2 = 3, k = 100)$ to illustrate the idea; we might begin with w_0 (often called the **seed**) equal to 7. Then

$$w_1 = 21 \times 7 + 3 \ (\text{mod } 100) = 147 + 3 \ (\text{mod } 100) = 150 \ (\text{mod } 100) = 50.$$
$$w_2 = 21 \times 50 + 3 \ (\text{mod } 100) = 1050 + 3 \ (\text{mod } 100) = 1053 \ (\text{mod } 100) = 53.$$

Division by the modulus 100 puts the numbers in the interval (0, 1) and we obtain the sequence: .07, .50, .53, .16, .39, ...

We can outline an algorithm to generate random numbers:

(1) input seed w_0, an integer;
(2) $w_1 = a_1 w_0 + a_2$ (mod k); the result is an integer;
(3) $u = w_1/k$, where the result is truncated to a given number of places;
(4) output u, a sample from U(0, 1).

The choice of a_1, a_2 and k is critical for the performance of the generators. Two general classes of generators, depending on whether $a_2 = 0$ or not, may be recognised. The first congruential generators were of the form

$$w_{i+1} = a_1 w_i \quad (\text{mod } k)$$

and, for obvious reasons, are called **multiplicative** congruential generators. The generators containing the additional positive integer a_2 are called **mixed** congruential generators. When choosing the parameters of a generator the properties that can be taken account of are (a) efficiency, (b) long cycle length, and (c) small serial correlation. Here efficiency means quick generation. Steps (2) and (3) of the algorithm above involve division, which is usually a relatively slow operation on a computer. Therefore it is desirable to choose the modulus k to make division by it quick and simple. This can be achieved if k is a power of the number base b of the computer (most commonly $b = 2$, but it may equal 10). The parallel in ordinary arithmetic is when, working with base 10, the division of 179236 by 10^3 is accomplished simply by moving a decimal point, to give 179.236. (Division by $b^p \pm 1$ is also achieved simply on a computer with number base b, by subtracting the appropriate number from the word (or bytes) containing the exponent.) We also want k to be as large as possible so that the maximum cycle is long, and hence efficiency suggests that the modulus k should equal b^l (or $b^l \pm 1$), where b is the number base of the computer and l is related to the word length of the computer. (For the common microcomputers an appropriate value of l is 15; with mainframe computers it will be about twice this.) We assume, of course, that the random number generator will be programmed in machine code.

Rules for the choice of parameters to give the maximum possible cycle length with mixed congruential generators have been found. A sequence from a mixed congruential generator has cycle length k if and only if:

(i) a_2 and k have no common factors except 1;
(ii) $(a_1 - 1)$ is a multiple of every prime that divides k;
(iii) $(a_1 - 1)$ is a multiple of 4 if k is a multiple of 4.

It is easy to see that, if k is chosen as recommended in the previous paragraph, then to obey these rules we should choose a_1 equal to $20c + 1$ (c being a positive integer) on a decimal machine, and equal to $4c + 1$ on a binary machine.

Maximum possible cycle lengths are not obtainable with multiplicative congruential generators, though long ones can be found. For example, if $k = 2^\alpha$ then a cycle length of $2^{\alpha-2}$ may be obtained. With these generators, the seed w_0 must be chosen so that it has no factors in common with the modulus k.

In theory, the sample values obtained are required to be independent, so the serial correlations will be zero. The most important requirement in practice is that the first serial correlation (the correlation of u_i with u_{i+1}) shall be small, for we cannot ensure that it is zero. It has been shown that with a mixed congruential generator in which a_2 is much less than k, this requirement can be satisfied by choosing a_1 approximately equal to $\sqrt{k}$.

Unfortunately, satisfying these desirable properties does not ensure that we have a good random number generator. The only way to check whether a generator is producing suitably 'random' numbers is to apply to the numbers the tests described in Section 7.6; but first we will give an example of a random number generator.

7.5   Examples of linear congruential random number generators

Ripley (1983) states that: 'Today the best-understood way to generate pseudo-random numbers is to use a congruential generator with k chosen as large as possible (and at least 2^{30}) and $a1$ and $a2$ chosen to ensure full or maximal period *and* good lattice behaviour in up to four dimensions.' (We have changed his symbols to agree with ours.) We describe briefly in Section 7.6.4 what is meant by good lattice behaviour; it is too complicated to discuss in detail. From a table in Ripley's paper we have chosen four generators that appear to satisfy his criteria except that, for a reason that will become clear later, we have included a generator with a k value of 2^{16} (well below his suggested lower bound of 2^{30}).

k	$a1$	$a2$
2^{59}	13^{13}	0
2^{32}	69069	odd
$2^{31}-1$	513	0
2^{16}	293	odd

The lattice properties of these generators improve as k increases.

As we pointed out in Section 7.4, random number generators are usually written in machine code. But we think the reader will find it helpful to see a congruential generator coded in Pascal (Algorithm 7.2) so that he may investigate in detail how a generator works. Algorithm 7.2, a function *Random*, is coded from the outline algorithm given in Section 7.4. It may be used to generate a sequence of random numbers in the interval (0, 1). When *Random* is being used in a program, declare a global integer variable *Seed* and initialise it before the first call of the function. The value of *Seed* changes as a result of each call of *Random*, so it must be passed as a variable-parameter. The initialisation may be done by asking the program user to input a value, or by making use of the time function (if available) of the computer.

In the algorithm, multiplication and calculation of the remainder, modulo 2^{16}, are carried out in integer arithmetic and must be done exactly. This may lead to a difficulty in programming in a high-level language. The maximum value taken by $(a1*Seed + a2)$ must not exceed the maximum integer allowed with the computer being used. That is why we have chosen $k = 2^{16}$ in our example; the higher values of k in the table are likely to give inadmissible integers on many computers.

```
function Random(var Seed : integer) : real;

(* used repeatedly generates a sequence of *)
(* independent samples from U(0,1).        *)

const a1 = 293;
      a2 = 1;
      a3 = 65536;
```

```
begin
    Seed := a1*Seed + a2;
    Seed := Seed mod a3;
    Random := Seed/a3;
end;  (* of Random *)
```

Algorithm 7.2 *Random*

Program 7.1, *CardCollection*, illustrates how a random number generator may be used in a simulation program. The function *Random* is called repeatedly, and the integer *Seed* generated by the last call of the function is used as argument for the next call.

7.6   Tests of randomness

The only reasonable guarantee we can have that a pseudo-random number generator will be satisfactory is that it has passed tests of the statistical properties of the numbers it produces. These properties may be summed up as the requirements of **uniformity** and **independence**. Many statistical tests have been used on random numbers; we have space for a few only. More tests are given in Fishman (1978) and Ripley (1983), for example. A very good general discussion is given in Knuth (1981).

7.6.1   Frequency test
This is a test of uniformity. If the (0, 1) interval is divided into a number of classes of equal width, then we would expect an equal number of random numbers to fall in each class. For example, if we use 10 intervals each of length 0.1 then, with 1000 random numbers, we would expect 100 to fall in each of the ten intervals (classes). The frequencies of the numbers falling in each class may be found using Algorithm 4.4, *FormFreqTable*. The chi-squared test of equal frequencies may be done using Algorithm 10.5, *FindX2ofFit*.

7.6.2   Serial test
We suppose the (0, 1) interval is divided into classes as in the frequency test. We then count the frequencies $f(i, j)$ of distinct pairs of random numbers (u_{2k-1}, u_{2k}), ($k = 1$, 2, . . . , n), such that the first of the pair falls in interval i and the second falls in interval j. These frequencies should all be equal if the numbers are uniform and independent. The chi-squared test may again be carried out with Algorithm 10.5, *FindX2ofFit*.

7.6.3   Serial correlation
The first serial correlation of the sequence $u_0, u_1, . . . , u_n$ is the correlation of u_i with u_{i+1}, where i runs through the values 0, 1, 2, . . . , $n-1$. The correlation coefficient may be calculated as described in Section 6.6, and tested for difference from zero as described in Exercises 10.7, question 2. The array x used in the algorithms will have elements $u_0, u_1, . . . u_{n-1}$ while the array y will have elements $u_1, u_2, . . . u_n$. Similarly, the tth serial correlation coefficient measures the correlation of u_i with u_{i+t}. These serial correlations, for all values of t, should be zero if the numbers being produced are independent.

7.6.4   Lattice Test

A method of investigating whether successive pseudo-random numbers act as independent random variables is to plot u_i against u_{i+1} for a complete period of the generator. There will be k points in the plot, assuming the congruence is modulo k. But the number of vertices that might be occupied is k^2. Thus the plotted points are relatively sparsely spread over the lattice of possible points. The sparseness is even greater if we consider plots of triples (u_i, u_{i+1}, u_{i+2}) in three dimensions or, more generally, of m-tuples in m dimensions. The lattice test attempts to quantify how sparse is the spacing of the observed m-tuples and so assess how 'independent' are the successive values of u: see Ripley (1983) for details.

[*Note*: We give no test data for the algorithms in this chapter, but some of the following exercises will serve this purpose.]

7.7   Exercises

1   Run Program 7.1, *CardCollection*, with *NoInSet* = 10. Repeat this a number of times, and observe the variation in the number of packets that must be bought to complete a set of cards.

Print the array *Frequency*, which contains the number of cards of each type that are collected in order to obtain a complete set.

2   We may call the interval between a card being obtained and another wanted card appearing a *waiting time*. In the *CardCollection* simulation this is measured as number of packets. Extend the program so that in any given simulation the waiting time to the ith card is stored in *WaitingTime*[i]; *Waiting Time*[1] will of course be 1. For each i (from 1 to *NoInSet*), total the waiting times for the k simulations; store these totals in *Total*. Then *Total*[i]/k will be the mean waiting time for card i; print this out for each i in addition to the other quantities printed by the program.

3   Generate 100 random numbers using Algorithm 7.2, *Random*. By using or adapting *FormFreqTable* (Algorithm 4.4) and *OutputFreqTable* (Algorithm 4.5), print out a frequency table using classes with lower limit 0 and class-width 0.1.

4   *Frequency test on pseudo-random numbers.* Construct a program to generate and test random numbers according to the description given in Section 7.6.1.

5   *Serial correlation in pseudo-random numbers.* Construct a program to generate random numbers, and to test them by calculating the first, second and third serial correlation coefficients (i.e. $t = 1, 2$ and 3) according to the description given in Section 7.6.3.

6   Simulate how the proportion of heads varies as a coin is tossed repeatedly, the probability of obtaining a head at each single toss being p. Print out the proportion of heads at intervals of 10 tosses.

7   Choose a random sample of 12 values from the uniform distribution on (0, 1) using *Random* (Algorithm 7.2). Add the 12 values and call the result y. Repeat 1000 times.
Find the mean and variance of the 1000 values of y and construct a frequency table of

the values. What theoretical distribution does the distribution of y resemble? (See Section 9.5.1).

8 Write a program to investigate the following problem. There are n people at a party; for what value of n is the probability that there are at least two people with the same birthday equal to (a) 0.5, (b) 0.95?

As a first stage, write a procedure which chooses up to n values at random from the integers $1, 2, \ldots, 365$ and sets a variable equal to 1 if a value repeats and to 0 if there is no repeat.

8 Probability functions of random variables

8.1 Introduction

There are two functions associated with each discrete random variable X: the **probability mass function** and the **cumulative distribution function** (*ABC*, Chapter 7). If we denote the probability mass function by the letter f, then the probability that X takes the particular value x is

$$\Pr(X = x) = f(x).$$

For example, in the binomial distribution having parameters n and p,

$$f(x) = \binom{n}{x} p^x (1-p)^{n-x}$$

where x may take any of the values 0, 1, 2, ..., n.

The letter F is used to denote the cumulative distribution function, which gives the probability that the random variable X is less than or equal to some particular value, say b. Thus

$$F(b) = \Pr(X \leq b).$$

For a random variable taking values 0, 1, 2, ..., n, such as the binomial,

$$F(b) = \sum_{x=0}^{b} f(x) \qquad (b \leq n).$$

Similarly, two functions are defined for each continuous random variable X (*ABC*, Chapter 13). The **probability density function** $f(x)$ of X is a function whose integral from $x = a$ to $x = b$ ($b \geq a$) gives the probability that X takes a value in the interval (a, b):

$$\Pr(a \leq X \leq b) = \int_a^b f(x)\mathrm{d}x.$$

The **cumulative distribution function** $F(b)$ is such that

$$F(b) = \Pr(X \leq b) = \int_{-\infty}^b f(x)\mathrm{d}x.$$

With the exponential distribution, for example, the probability density function is

$$f(x) = \lambda \mathrm{e}^{-\lambda x} \qquad (x \geqslant 0, \lambda > 0),$$

and its cumulative distribution function is

$$F(b) = \int_0^b \lambda e^{-\lambda x}\,dx = 1 - e^{-\lambda b} \qquad (b \geq 0).$$

In practical statistics, two common requirements are to calculate **probabilities** and **percentage points** associated with theoretical distribution functions. For example, we may wish to calculate the probability that an item sampled at random from a normal distribution takes a value less than 1.5. This probability is 0.93 and we obtain it by evaluating the cumulative distribution function $F(1.5)$. More generally, we determine for a given value x_p of the variable the probability $p = F(x_p)$. Alternatively, we may wish to find the value of a normal random variable below which exactly 95% of randomly sampled items will fall. The value is 1.96, and it is called the 95 percentage point of the distribution. The general problem is, for given p, to solve the equation $F(x_p) = p$. We may think of the process in terms of a new function, the inverse distribution function, which produces x_p given p.

In this chapter we consider how to calculate the values of (i) probability mass and density functions; (ii) distribution functions ('probabilities'); and (iii) inverse distribution functions ('percentage points'), for common statistical distributions.

8.2 Discrete random variables

8.2.1 Binomial distribution
The probability of x successes in n independent trials, when the probability of success at any trial is p, is given by

$$f(x) = \binom{n}{x} p^x q^{n-x} \qquad (x = 0, 1, \ldots, n),$$

where $q = 1 - p$. The function *Binomial* (Algorithm 8.1) is based on the recurrence formula relating the probability of i successes $P(i)$ to the probability of $i - 1$ successes, i.e.

$$P(i) = \frac{(n - i + 1)}{i} \frac{p}{q} P(i - 1).$$

with the starting value $P(0) = q^n$.

```
function Binomial(x,n:integer;p:real):real;

(* calculates probability of x successes in n independent *)
(* trials when probability of success at each trial is p. *)

var i:integer; (* loop control *)
    nPlusOne:integer;
    q:real; (* = 1 - p *)
    Term:real; (* binomial term *)
    w:real; (*temporary store for finding n-th power of q *)
```

```
begin
    q:=1-p;w:=1;
    for i:=1 to n do
        w:=q*w;
    Term:=w; (* n-th power of q *)
    i:=0;nPlusOne:=n+1;
    while i<x do
    begin
        i:=i+1;
        Term:=(nPlusOne-i)*p*Term/(i*q)
    end;
    Binomial:=Term
end; (* of Binomial *)
```

Algorithm 8.1 *Binomial*

The jth element $(j = 0, 1, \ldots, n)$ of the array giving the values of the cumulative distribution function is $\sum_{i=0}^{j} P(i)$, so this array can be found by a very simple addition to Algorithm 8.1.

8.2.2 Poisson distribution
The probability mass function of the Poisson distribution with parameter λ (> 0) is given (*ABC*, Chapter 19) by

$$f(x) = \frac{\lambda^x}{x!} e^{-\lambda} \qquad (x = 0, 1, 2, \ldots).$$

The calculation of the probabilities may be carried out in a similar way to that for the binomial distribution (Algorithm 8.1), but based on the recurrence relation

$$P(i) = \lambda P(i-1)/i \qquad (i = 1, 2, 3, \ldots)$$

where $P(0) = e^{-\lambda}$.

8.2.3 Geometric distribution
The probability mass function of the geometric distribution with parameter p is given (*ABC*, Section 7.7) by

$$f(x) = p(1-p)^{x-1} \qquad (x = 1, 2, \ldots).$$

An appropriate recurrence relation is

$$P(i) = (1-p)P(i-1),$$

i.e. $\qquad P(i) = qP(i-1),$

where $q = 1 - p$, for $i = 2, 3, 4, \ldots$; and

$$P(1) = p.$$

8.3 The exponential distribution

The probability density function of the exponential distribution, which is a continuous random variable, is given (*ABC*, Section 13.6) by

$$f(x) = \lambda e^{-\lambda x} \qquad (x \geq 0),$$

where $\lambda > 0$. The distribution has mean $1/\lambda$. The cumulative distribution function is given by

$$F(b) = 1 - e^{-\lambda b}.$$

One line of program will serve to calculate each of these functions for a particular value of x (or b). Thus

$$Exponential := l * \exp(-l * x)$$
$$ExponentialProb := 1 - \exp(-l * x).$$

Each must of course be set within an appropriate function module with input parameters l and x.

Sometimes we wish to use the inverse distribution function; that is, to calculate the value of x corresponding to a particular value p of the cumulative distribution function. We therefore have

$$p = 1 - e^{-\lambda x}$$

whose solution is $x = -\lambda^{-1} \ln(1-p)$. An appropriate instruction would be

$$ExponentialPercentPoint := -\ln(1-p)/l.$$

8.4 Calculation of integrals

The probability density functions of most of the common continuous random variables look more complicated than that of the exponential distribution, but evaluation of them by computer is straightforward. Evaluating the cumulative distribution functions may be awkward since there is not always an equivalent explicit expression. The most direct approach in such cases is to use numerical integration of the probability density function. We therefore give Algorithm 8.2, *SimpsonIntegral*, for integration using Simpson's rule. The probability density function being integrated is a function f.

In evaluating a cumulative distribution function, the lower limit of the necessary integral is the lower limit of the possible values taken by the random variable: the range of integration begins at the left-hand end or tail of the distribution and moves up to the particular value b of interest. (Sometimes it is easier to find an integral related to this required one.)

The accuracy of the value of the integral we obtain by this method depends upon the number n of intervals into which the range of integration (a, b) is divided and the 'smoothness' of the integrand. Numerical analysis gives us a bound (upper limit) for the error which is

$$\varepsilon = \left| \frac{M(b-a)^5}{2800\,n^4} \right|$$

where M is the maximum value of the modulus of the fourth derivative of the integrand in the interval. One approach is to substitute for ε the accuracy within which we wish to work, and calculate an appropriate n. We follow the method given by Wirth (1973). The calculation is repeated, with the number of intervals doubling at each repeat, until two successive values of the calculated integral do not change at a given level of accuracy (more precisely, the difference between two successive values is less than the declared 'error'). On the kth evaluation ($k = 1, 2, \ldots$) the interval length h is $(b-a)/n$, where n equals 2^k. The function f is evaluated at the endpoints of the n intervals to give the function values $\{f_i; i = 0, 1, \ldots, n\}$, where $f_i = f(a + i*h)$. By Simpson's Rule an approximation to the value of the integral $\int_a^b f(x)\,dx$ is

$$\frac{h}{3}(f_0 + 4f_1 + 2f_2 + 4f_3 + 2f_4 + \ldots + 4f_{n-3} + 2f_{n-2} + 4f_{n-1} + f_n).$$

The sum in brackets may be considered to be made up of three parts and written

$$S^{(1)} + 2S^{(2)} + 4S^{(4)}$$

where $S^{(i)}$, with $i = 1, 2$ or 4, is the sum of the function values which have coefficient i.

```
function SimpsonIntegral(function f(t:real) : real;
                         a,b : real) : real;

(* integrates the function f from lower limit a to upper *)
(* limit b choosing an interval length so that the error *)
(* is less than a given amount  -  default value 1.0e-6. *)

const Error = 1.0e-6

var h : real; (* current length of interval *)
    i : integer; (* counter *)
    Integral : real; (* current approximation to integral *)
    LastInt : real; (* previous approximation *)
    n : integer; (* number of intervals *)
    Sum1,Sum2,Sum4 : real; (* sums of function values *)
begin
    n := 2;h := 0.5*(b - a);
    Sum1 := h*(f(a) + f(b));
    Sum2 := 0;
    Sum4 := f(0.5*(a + b));
    Integral := h*(Sum1 + 4*Sum4);
    repeat
        LastInt := Integral; n := n + n; h := 0.5*h;
        Sum2 := Sum2 + Sum4;
        Sum4 := 0; i := 1;
        repeat
            Sum4 := Sum4 + f(a + i*h);
            i := i + 2
        until i > n;
        Integral := h*(Sum1 + 2*Sum2 + 4*Sum4);
    until abs(Integral - LastInt) < Error;
    SimpsonIntegral := Integral/3
end; (* of SimpsonIntegral *)
```

Algorithm 8.2 *SimpsonIntegral*

Many of the function values in the kth evaluation are the same as those in the $(k-1)$th evaluation. We may use this fact to make the computation more efficient. Writing $S_k^{(i)}$ for the sum of the function values with coefficient i in the kth evaluation, we find that

$$S_k^{(1)} = S_{k-1}^{(1)}; \qquad S_k^{(2)} = S_{k-1}^{(2)} + S_{k-1}^{(4)}.$$

The function values with coefficient 4 in the kth evaluation appear for the first time in that evaluation, and

$$S_k^{(4)} = f(a+h) + f(a+3h) + \ldots + f(a+(n-1)h).$$

The initial values of the sums are

$$S_1^{(1)} = f(a) + f(b); \qquad S_1^{(2)} = 0; \qquad S_1^{(4)} = f((a+b)/2).$$

Algorithm 8.2 follows from this description.

8.5 The normal integral

8.5.1 Simpson's rule

The probability density function of the standard normal distribution (*ABC*, Section 14.2) is programmed in *Normal* (Algorithm 8.3). The distribution function, for argument z, is obtained by integrating the density from $-\infty$ to z.

```
function Normal(z : real) : real;

(* the density function of the standard normal distribution *)

const   a = 0.39894228; (* 1/sqrt(2*pi) *)

begin
    Normal := a*exp(-0.5*z*z)
end; (* of Normal *)
```
Algorithm 8.3 *Normal*

We cannot integrate over an infinite range using Simpson's rule, so we must rewrite the integral; in doing so we make use of the fact that the integrand is symmetrical about zero. We obtain

$$F(z) = 0.5 + \int_0^z \frac{1}{\sqrt{2\pi}} \exp\left(-\frac{t^2}{2}\right) dt.$$

This is programmed in *NormalProb* (Algorithm 8.4).

```
function NormalProb(z : real) : real;

(* the distribution function of the standard normal        *)
(* distribution derived by integration using Simpson's rule *)

begin
    NormalProb := 0.5+SimpsonIntegral(Normal,0.0,z);
end; (* of NormalProb *)
```
Algorithm 8.4 *NormalProb*

8.5.2   An approximation to normal probabilities

An alternative approach is to use an approximation that is an explicit function of z. We give an approximation due to Hastings; it is quoted in Abramowitz and Stegun (1972), where more accurate—though more complicated—approximations are also quoted. For non-negative z,

$$F(z) = 1 - 0.5(1 + a_1 z + a_2 z^2 + a_3 z^3 + a_4 z^4)^{-4} + \varepsilon(z).$$

The values of constants $a_1, \ldots, a_4$ are given in Algorithm 8.5, *NormalProbApprox*.

```
function NormalProbApprox (z : real) : real;

(* finds, using rational approximation, the distribution *)
(* function of the standard normal distribution.         *)

const a1 = 0.196854;
      a2 = 0.115194;
      a3 = 0.000344;
      a4 = 0.019527;
var   w : real; (* stores intermediate results *)

begin
    w := abs(z);
    w := 1+w*(a1+w*(a2+w*(a3+w*a4)));
    w := w*w*w*w;
    w := 1-0.5/w;
    if z >= 0
        then NormalProbApprox := w
        else NormalProbApprox := 1-w;
end; (* of NormalProbApprox *)
```

Algorithm 8.5 *NormalProbApprox*

We evaluate the expression using the absolute value of z and obtain the value $F(|z|)$. This is the correct value for $z \geq 0$; for $z < 0$ we require $1 - F(|z|)$. The error of the approximation is $\varepsilon(z)$, and it is stated that $|\varepsilon(z)| < 2.5 \times 10^{-4}$. There are also, of course, rounding errors in the evaluation.

8.5.3   Normal percentage points by approximation

The cumulative distribution function $F(z)$ gives the probability that a value chosen at random from the distribution will be less than z. Sometimes we wish to carry out the inverse operation: given the probability, we wish to find the corresponding percentage point z. Provided that $0.5 \leq p < 1$ an approximate value of z corresponding to a particular value of p is given by

$$z = t - \frac{a_1 + a_2 t}{1 + a_3 t + a_4 t^2} + \varepsilon(p),$$

where $t = \sqrt{-2 \ln(1 - p)}$. The absolute error $|\varepsilon(p)| < 3 \times 10^{-3}$. The limitation on the range of p is removed if we work with $p' = 0.5 + \text{abs}(p - 0.5)$; the answer is correct if $p \geq 0.5$ and must be given a negative sign if $p < 0.5$.

We give Algorithm 8.6, *NormalPercentPointApprox*, to carry out this calculation.

```
function NormalPercentPointApprox(p:real):real;

(* finds the value of z such that F(z)=p, where F is  *)
(* the standard normal distribution function.         *)

const a1=2.30753;
      a2=0.27061;
      a3=0.99229;
      a4=0.04481;

var p1:real; (* adjusted p *)
    t:real; (* argument of function *)
    w:real; (* stores intermediate resuls of evaluation *)

begin
    p1:=0.5+abs(p-0.5);
    t:=sqrt(-2.0*ln(1-p1));
    w:=a1+a2*t;
    w:=w/(1.0+t*(a3+a4*t));
    w:=t-w;
    if p>=0.5
        then NormalpercentPointApprox:=w
        else NormalPercentPointApprox:=-w;
end; (* of NormalPercentPointApprox *)
```

Algorithm 8.6 *NormalPercentPointApprox*

8.6   Distributions based on the normal distribution

The sampling distributions t, χ^2 and F are important in both theoretical and practical statistics. Their probability density functions are as follows.

i) t with k degrees of freedom:

$$\left(\frac{\Gamma[(k+1)/2]}{\sqrt{k\pi}\,\Gamma[k/2]}\right)\left(\frac{1}{(1+x^2/k)^{(k+1)/2}}\right) \qquad -\infty < x < \infty.$$

ii) χ^2 with k degrees of freedom:

$$\frac{1}{\Gamma[k/2]2^{k/2}}\,x^{k/2-1}\,e^{-x/2} \qquad 0 < x < \infty.$$

iii) F with k_1 and k_2 degrees of freedom:

$$\left(\frac{\Gamma[(k_1+k_2)/2](k_1/k_2)^{k_1/2}}{\Gamma[k_1/2]\Gamma[k_2/2]}\right)\left(\frac{x^{(k_1-2)/2}}{(1+k_1x/k_2)^{(k_1+k_2)/2}}\right) \qquad 0 < x < \infty.$$

These distributions are defined in terms of normal variables in Chapter 9, and their density functions are derived in Hogg and Craig (1970). Their uses are discussed in *ABC* (Chapters 16 and following).

The density functions involve the gamma function $\Gamma(x)$, which is closely related to factorials. The function $\Gamma(x)$ is equal to $(x-1)\Gamma(x-1)$, and hence if we put x equal to a positive integer n we obtain $\Gamma(n) = (n-1)!$ (since $\Gamma(1) = 1$). In the density functions above, the arguments are all such that twice the argument is a positive integer. Using the relation between $\Gamma(x)$ and $\Gamma(x-1)$, it follows that, if n is a positive integer, $\Gamma(n+\frac{1}{2})$ equals $(n-\frac{1}{2})(n-\frac{3}{2})(n-\frac{5}{2}) \ldots \frac{3}{2} \cdot \frac{1}{2} \Gamma(\frac{1}{2})$; it can be shown also that $\Gamma(\frac{1}{2}) = \sqrt{\pi}$.

Now the gamma function can be dangerous to use on a computer since it very quickly overflows: an argument of 35 or 36 saturates it. We therefore work with $\ln \Gamma(x)$ instead, since we shall find this easy to incorporate into later algorithms. Algorithm 8.7, *LnGamma*, evaluates the logarithm of the function for arguments of the form w, where $2w$ is a positive integer.

```
function LnGamma(w : real) : real;

(* calculates the logarithm of the gamma function; *)
(* w must be such that 2*w is an integer > 0.      *)

const a = 0.57236494; (* ln(sqrt(pi)) *)
var Sum:real; (* temporary store for summation of values *)

begin
    Sum := 0;
    w := w-1;
    while w > 0.0 do
    begin
        Sum := Sum + ln(w);
        w := w-1
    end; (* of summation loop *)
    if w<0.0
        then LnGamma := Sum + a
        else LnGamma := Sum
end; (* of LnGamma *)
```

Algorithm 8.7 *LnGamma*

The cumulative distribution functions may be evaluated using Simpson's rule, as was done in Algorithm 8.4 for the normal distribution. Serious difficulties arise in the integration for χ^2 with one degree of freedom and for F when $k_1 = 1$. In both these cases the integrand is singular at the origin, and the number of strips required in the integration is prohibitively large even to obtain very approximate results.

8.6.1   The *t* distribution
An exact algorithm (see Abramowitz and Stegun (1972)) for finding the probability p that corresponds to a given value of t with k degrees of freedom can be constructed from a series summation. If we define $\theta = \tan^{-1}(t/\sqrt{k})$, then $p = \frac{1}{2}(1 + A)$, where

$$A = 2\theta/\pi \qquad (k = 1),$$

$$A = \frac{2}{\pi}\left[\theta + \sin\theta\left(\cos\theta + \frac{2}{3}\cos^3\theta + \ldots + \frac{2.4 \ldots (k-3)}{1.3 \ldots (k-2)}\cos^{k-2}\theta\right)\right]$$

$$(k > 1 \text{ and odd}),$$

$$A = \sin\theta \left[1 + \tfrac{1}{2}\cos^2\theta + \frac{1.3}{2.4}\cos^4\theta + \ldots + \frac{1.3\ldots(k-3)}{2.4\ldots(k-2)}\cos^{k-2}\theta \right]$$

$$(k \text{ even}).$$

Algorithm 8.8, *tProb*, carries out this summation.

```
function tProb(t:real; k:integer) : real;

(* the distribution function of the t distribution for *)
(* k degrees of freedom                                *)

const a=0.636619772; (* 2/pi *)
var c,s:real; (* cosine and sine of theta *)
    i:integer; (* counter for terms of series *)
    Sum:real; (* temporary store for summing series *)
    Term:real; (* term of series *)
    Theta:real; (* argument of terms *)

begin
    Term := k;
    Theta := arctan(t/sqrt(Term));
    s := sin(Theta);
    c := cos(Theta);
    Sum := 0.0;
    if k > 1
    then begin
        if odd(k)
        then begin
            i := 3;
            Term := c
        end
        else begin
            i := 2;
            Term := 1.0
        end;
        Sum := Term;
        while i < k do
        begin (* summation of series *)
            Term := Term*c*c*(i-1)/i;
            Sum := Sum + Term;
            i := i + 2
        end; (* of while *)
        Sum := s*Sum
    end;
    if odd(k) then Sum := a*(Sum + Theta);
    tProb := 0.5*(1.0+Sum)
end; (* of tProb *)
```

Algorithm 8.8 *tProb*

8.6.2 The χ^2 distribution
The algorithm that we give for the χ^2 distribution function is based on an algorithm by Lau (1980) for the gamma distribution function. The gamma distribution with parameters α, β (usually denoted by $G(\alpha, \beta)$) has probability density function (Hogg and

Craig, 1970):

$$\frac{1}{\Gamma(\alpha)\beta^\alpha} x^{\alpha-1} e^{-x/\beta} \qquad 0 < x < \infty; \qquad \alpha, \beta > 0.$$

The parameter β is called a **scale parameter** since a transformation $y = cx$ changes only the value of β in the density function; α is called a **shape parameter**. The χ^2 distribution with k degrees of freedom is the gamma distribution $G(k/2, 2)$, as may be seen by comparing the probability density functions. In Algorithm 8.9, *ChiSquaredProb*, the χ^2 distribution input values are modified in the first two operative statements and pass into an algorithm for the gamma distribution function. The algorithm after the change of input values may be treated as one for the distribution function of $G(k_1, 1)$, with input value x.

```
function ChiSquaredProb(x:real ; k:integer) : real;

(* the distribution function of the chi-squared distribution *)
(* based on k degrees of freedom. Uses LnGamma.              *)

var Factor : real; (* factor which multiplies sum of series *)
    g : real; (* LnGamma(k1+1) *)
    k1 : real; (* adjusted degrees of freedom *)
    Sum : real; (* temporary storage for partial sums *)
    Term : real; (* term of series *)
    x1 : real; (* adjusted argument of function *)

begin
    x1 := 0.5*x;
    k1 := 0.5*k;
    g := LnGamma(k1+1);
    Factor := exp(k1*ln(x1)-g-x1);
    Sum := 0;
    if Factor > 0
    then begin
        Term := 1; Sum := 1;
        while Term/Sum > 1E-6 do
        begin
            k1 := k1 + 1;
            Term := Term*x1/k1;
            Sum := Sum + Term
        end (* of while *)
    end; (* of if-then loop *)
    ChiSquaredProb := Sum*Factor
end; (* of ChiSquaredProb *)
```

Algorithm 8.9 *ChiSquaredProb*

8.6.3 The *F* distribution
We give the algorithm *FProb* (8.10). It uses an algorithm for the incomplete beta function:

$$I_x(a, b) = \frac{\Gamma(a)\Gamma(b)}{\Gamma(a+b)} \int_0^x t^{a-1} (1-t)^{b-1} \, dt,$$

given by Majumder and Bhattacharjee (1973*a*); the underlying method is to sum a series of integrations by parts. The algorithm (*BetaRatio*) is rather long, so we print it in the Appendix (A.1, page 153). The integral $I_x(a, b)$ equals the value of the F distribution function, with k_1 and k_2 degrees of freedom and with argument F when $a = k_2/2$, $b = k_1/2$ and $x = k_2/(k_2 + k_1 F)$.

```
function FProb(f : real; k1,k2 : integer) : real;

(* the distribution function of the F distribution based on *)
(* k1 and k2 degrees of freedom;  uses functions LnGamma    *)
(* and BetaRatio.                                           *)

var h1,h2 : real; (* modified degrees of freedom *)
    LnBeta : real; (* log of complete beta function with *)
                   (* parameters h1 and h2.              *)
    x : real; (* argument of incomplete beta function *)


function LnGamma(w : real) : real;

        (*** Algorithm 8.7 ***)


function BetaRatio(x,a,b,LnBeta : real) : real;

        (*** Algorithm A.1 ***)

begin    (* FProb *)
    h1 := 0.5*k1;
    h2 := 0.5*k2;
    x := h2/(h2+h1*f);
    LnBeta := LnGamma(h1) + LnGamma(h2) - LnGamma(h1+h2);
    FProb := 1 - BetaRatio(x,h2,h1,LnBeta)
end; (* of FProb *)
```

Algorithm 8.10 *FProb*

8.6.4   Approximations to distribution functions of t, χ^2 and F

For many purposes an approximation to the value of the distribution function may be adequate. We give below transformations of t, χ^2 and F that will produce, in each case, a variable z which is approximately a standard normal variable (i.e. one with mean 0 and variance 1). Thus for a given value of t, say, we can derive the corresponding value of z and then use an algorithm to find the value of the distribution function corresponding to z, and hence to the original t value.

Wallace (1959) proposed for t with k degrees of freedom the transformation

$$z = \frac{8k+1}{8k+3}\left[k\ln\left(1+\frac{t^2}{k}\right)\right]^{1/2}.$$

For χ^2 with k degrees of freedom, Wilson and Hilferty (1931) showed that $(\chi^2/k)^{1/3}$ is approximately normal with mean $\left(1 - \dfrac{2}{9k}\right)$ and variance $\left(\dfrac{2}{9k}\right)$. Hence

$$z = \frac{(\chi^2/k)^{1/3} - (1 - 2/(9k))}{(2/(9k))^{1/2}}.$$

This approximation may be applied to derive one for F. The F distribution with k_1 and k_2 degrees of freedom, $F_{(k_1, k_2)}$, is defined as the ratio (apart from a factor) of two χ^2 variables (Section 9.6). Thus we may write

$$F = (\chi_1^2/k_1)/(\chi_2^2/k_2),$$

the chi-squared distributions χ_1^2 and χ_2^2 having respectively k_1 and k_2 degrees of freedom. Hence

$$F^{1/3} = (\chi_1^2/k_1)^{1/3}/(\chi_2^2/k_2)^{1/3}$$

is the ratio of two variables which are approximately normally distributed. There is a result due to Geary (1930) concerning the ratio $v = z_1/z_2$, where z_1 and z_2 are normal variables with means μ_1, μ_2 and variances σ_1^2, σ_2^2 respectively: he showed that

$$z = \frac{\mu_1 - \mu_2 v}{(\sigma_1^2 + \sigma_2^2 v^2)^{1/2}}$$

is approximately a standard normal variable. If we substitute $F^{1/3}$ for v, and the appropriate values for μ_1, μ_2, σ_1^2, σ_2^2, we obtain

$$z = \frac{\left(1 - \dfrac{2}{9k_2}\right)F^{1/3} - \left(1 - \dfrac{2}{9k_1}\right)}{\left(\dfrac{2}{9k_2}F^{2/3} + \dfrac{2}{9k_1}\right)^{1/2}}.$$

Paulson (1942) appears to have been the first person to propose this. Algorithm 8.11, *FProbApprox*, makes use of this result. It uses *NormalProbApprox* (Algorithm 8.5) as a subroutine. Algorithm 8.11 is valid only for values of $F \geq 1$. If $F < 1$, then F should be replaced by $1/F$, k_1 and k_2 interchanged, and the resulting probability subtracted from 1. A correction to the z value is required in the algorithm when $k_2 \leq 3$ in order to improve accuracy for these small values of k_2. With this correction incorporated, Algorithm 8.11 gives reasonably satisfactory results; for example it calculates upper tail probabilities in the region of 0.05 with a relative accuracy of a few percent over the whole range of values of k_1, k_2.

We can also use this algorithm to derive approximate values of the distribution functions of t and χ^2, using the following relations:

$$t_{(k)}^2 = F_{(1, k)}$$

$$F(t) = \tfrac{1}{2}(1 - F(F)) \qquad t < 0$$
$$F(t) = \tfrac{1}{2}(1 + F(F)) \qquad t > 0$$

and $\chi_{(k)}^2/k = F_{(k, \infty)},$

where k denotes degrees of freedom.

Thus to find the value of the distribution function of t corresponding to a given value x, enter x^2 into Algorithm 8.11 using 1 and k as degrees of freedom; we obtain from the

```
function FProbApprox (f : real; k1,k2 : integer ) : real;

(* calculates the F distribution function using Paulson's *)
(* approximation.  Uses NormalProbApprox.                 *)

var a1,a2 : real; (* coefficients *)
    w1,w2 : real; (* numerator and square of denominator *)
    fThird : real; (* cube root of f *)
    z : real; (* normal value corresponding to f *)
begin
    a1 := 2.0/(9.0*k1);
    a2 := 2.0/(9.0*k2);
    fThird := exp(ln(f)/3.0);
    w1 := fThird-a2*fThird+a1-1;
    w2 := a2*fThird*fThird+a1;
    z := w1/sqrt(w2);
    (* improved approximation for small k2 *)
    if k2<4 then z := z*(1.0+0.08*z*z*z*z/(k2*k2*k2));
    FProbApprox := NormalProbApprox(z)
end; (* of FProbApprox *)
```

Algorithm 8.11 *FProbApprox*

algorithm an approximate value of the required probability. When dealing with χ^2 we cannot of course put $k_2 = \infty$, but using a very large number for k_2 will be satisfactory. As the reader will see, it is possible by using these relations to produce a relatively short algorithm to give probabilities for normal, t, χ^2 and F variables.

8.6.5 Percentage points based on approximations

Approximate values of the percentage points of the sampling distributions can be found conveniently using the approximations of Section 8.6.4. We may find, for example, the value of t which corresponds to the value 0.95 of the distribution function.

If t has k degrees of freedom, the approximation of Section 8.6.4 can be rewritten

$$t_{(k)} = \left[k \left(\exp \left(w^2/k\right) - 1\right)\right]^{1/2},$$

in which $w = z(8k + 3)/(8k + 1)$. Algorithm 8.12, *tPercentPointApprox*, uses this

```
function tPercentPointApprox( p : real; k : integer) : real;

(* calculates, using Wallace's approximation, the inverse t *)
(* distribution function based on k degrees of freedom.     *)
(* Uses NormalPercentPointApprox.                           *)

var w : real; (* temporary variable *)
    z : real; (* normal value corresponding to p *)

begin
    z := NormalPercentPointApprox(p);
    w := z*(1.0+2.0/(1.0+8.0*k));
    w := k*(exp(w*w/k)-1.0);
    tPercentPointApprox := sqrt(w)
end; (* of tPercentPointApprox *)
```

Algorithm 8.12 *tPercentPointApprox*

relation; it requires *NormalPercentPointApprox* (Algorithm 8.6) as a function. If we input the probability *p* and the degrees of freedom k_1, Algorithm 8.12 calculates first the corresponding standard normal variable *z*, and then goes on to calculate the percentage point of the *t* distribution using the relation above.

The *t* values derived from Algorithm 8.12 with input probabilities *p* of .975 and .995, which correspond to two-tail *t* tests at 5% and 1% significance levels, are compared with the exact values in Table 8.1. The algorithm is not reliable for fewer than four degrees of freedom.

Table 8.1

Degrees of freedom		1	2	5	10	20	30	60	120	∞
$p = .975$	calculated	17.6	4.48	2.58	2.23	2.09	2.04	2.00	1.98	1.96
	exact	12.7	4.30	2.57	2.23	2.09	2.04	2.00	1.98	1.96
$p = .995$	calculated	142.9	11.17	4.07	3.18	2.85	2.75	2.66	2.62	2.58
	exact	63.6	9.92	4.03	3.17	2.84	2.75	2.66	2.62	2.58

In a similar way, we may obtain χ_k^2 in terms of *z*:

$$\chi_k^2 = k\left(1 - \frac{2}{9k} + z\sqrt{\frac{2}{9k}}\right)^3.$$

Algorithm 8.13, *ChiSquaredPercentPointApprox*, uses this equation. Calculated and exact χ^2 values corresponding to probabilities *p* of .95 and .99 are given in Table 8.2.

```
function ChiSquaredPercentPointApprox(p:real; k:integer):real;

(* finds, using Wilson and Hilferty's approximation, the    *)
(* inverse chi-squared distribution function based on k      *)
(* degrees of freedom. Uses NormalPercentPointApprox.        *)

var a1 : real; (* coefficient *)
    w  : real; (* temporary store when evaluating expression *)
    z  : real; (* normal value corresponding to p *)

begin
    z  := NormalPercentPointApprox(p);
    a1 := 2.0/(9.0*k);
    w  := 1.0-a1+z*sqrt(a1);
    ChiSquaredPercentPointApprox := k*w*w*w
end; (* of ChiSquaredPercentPointApprox *)
```

Algorithm 8.13 *ChiSquaredPercentPointApprox*

The relation between *z* and *F* in Section 8.6.4 yields a quadratic in $F^{1/3}$, when a particular value is substituted for *z*. One root is positive and the other negative; the negative root is rejected. This calculation is carried out in Algorithm 8.14, *FPercentPointApprox*.

Table 8.2

Degrees of freedom		1	2	5	10	20	30	60
$p = .95$	calculated	3.75	5.94	11.04	18.29	31.4	43.8	79.1
	exact	3.84	5.99	11.07	18.31	31.4	43.8	79.1
$p = .99$	calculated	6.59	9.23	15.13	23.2	37.6	50.9	88.4
	exact	6.63	9.21	15.09	23.2	37.6	50.9	88.4

```
function FPercentPointApprox( p : real; k1,k2 : integer):real;

(* calculates,using Paulson's approximation, the inverse F *)
(* distribution based on k1 and k2 degrees of freedom.     *)
(* Uses NormalpercentPointApprox.                          *)

var a1,a2 : real; (* coefficients in expression *)
    u : real; (* temporary variable used when k2 is small *)
    w : real; (* positive root of quadratic *)
    w1,w2,w3 : real; (* coefficients of quadratic *)
    z : real; (* normal value corresponding to p *)
begin
    a1 := 2.0/(9.0*k1);
    a2 := 2.0/(9.0*k2);
    z := NormalPercentPointApprox(p);
    if k2 <= 3
    then begin
        u := sqrt(sqrt(k2));
        u := u*u*u;
        u := z/u;
        z := z*(1.1581-u*(0.2296-u*(0.0042+0.0027*u)));
    end;
    w1 := 1.0+a2*(a2-z*z-2);
    w2 := a1+a2-a1*a2-1;
    w3 := 1.0+a1*(a1-z*z-2);
    w := (-w2+sqrt(w2*w2-w1*w3))/w1;
    FPercentPointApprox := w*w*w
end; (* of FPercentPointApprox *)
```

Algorithm 8.14 *FPercentPointApprox*

The basic algorithm is not reliable when k_2 is less than four. Ashby (1968) proposed improving the value F given by the approximation, for $k_2 \leq 10$, by using a linear function $mF + c$ where m and c are constants depending on the significance level and on k_2 (but not on k_1). He gives a table of values of m and c for probabilities of .95, .99 and .999. No easy functional representation of the values appears to be possible so they have to be stored as a table in the computer. A substantial improvement in Algorithm 8.14, for $k_2 \leq 3$, can be obtained by replacing the value of z obtained from *NormalPercentPointApprox* with

$$z' = z(1.1581 - 0.2296u + 0.0042u^2 - 0.0027u^3)$$

where $u = z/k_2^{3/4}$. This correction is incorporated in Algorithm 8.14. Table 8.3 gives calculated and exact F values corresponding to probabilities p of .95 and .99.

Table 8.3 Comparison of *F* values calculated by *FPercentPointApprox* (Algorithm 8.14) with exact values. The *F* distribution has k_1 and k_2 degrees of freedom; *p* is the probability for which the *F* value is required.

$p = .95$

k_2	k_1	1	2	5	10	30	60
2	calculated	17.20	18.25	18.83	19.01	19.13	19.16
	exact	18.51	19.00	19.30	19.40	19.46	19.48
5	calculated	6.56	5.82	5.11	4.80	4.56	4.50
	exact	6.61	5.79	5.05	4.74	4.50	4.43
30	calculated	4.07	3.29	2.53	2.16	1.84	1.74
	exact	4.17	3.32	2.53	2.16	1.84	1.74

$p = .99$

k_2	k_1	1	2	5	10	30	60
2	calculated	71.31	78.85	83.97	85.79	87.04	87.36
	exact	98.50	99.00	99.30	99.40	99.47	99.48
5	calculated	16.76	14.01	11.81	10.93	10.29	10.13
	exact	16.26	13.27	10.97	10.05	9.38	9.20
30	calculated	7.50	5.39	3.71	2.98	2.39	2.21
	exact	7.56	5.39	3.70	2.98	2.39	2.21

8.6.6   Percentage points of the *F* distribution

To calculate a more precise value of *F* corresponding to a particular probability requires the solution of the equation

$$F(x) = p,$$

where $F(x)$ is the distribution function of *F* and *p* is the chosen probability. The solution may be obtained using Newton iteration; we give an appropriate algorithm (8.15), *FPercentPoint*, based on Majumder and Bhattacharjee (1973*b*). It uses Algorithms A.1, *BetaRatio*, and A.2, *InverseBetaRatio* (pages 153, 154). We have found the remarks of Cran *et al* (1977) helpful.

```
function FPercentPoint(p : real; k1,k2 : integer) : real;

(* calculates the inverse F distribution function based on *)
(* k1 and k2 degrees of freedom. Uses functions LnGamma,  *)
(* BetaRatio and InverseBetaRatio.                        *)

var h1,h2 : real; (* half degrees of freedom k1,k2 *)
    LnBeta : real; (* log of complete beta function with *)
                   (* parameters   h1 and h2            *)
    Ratio : real; (* Beta ratio *)
    x : real; (* inverse Beta ratio *)
```

```
function LnGamma(w : real) : real;

        (***   Algorithm 8.7  ***)



function BetaRatio(x,a,b,LnBeta : real) : real;

        (***   Algorithm A.1  ***)



function InverseBetaRatio(Ratio,a,b,LnBeta : real) : real;

        (***   Algorithm A.2  ***)
begin     (* FPercentPoint *)
    h1 := 0.5*k2;
    h2 := 0.5*k1;
    Ratio := 1 - p;
    LnBeta := LnGamma(h1) + LnGamma(h2) - LnGamma(h1+h2);
    x := InverseBetaRatio(Ratio,h1,h2,LnBeta);
    FPercentPoint := k2*(1 - x)/(k1*x)
end; (* of FPercentPoint *)
```

Algorithm 8.15 *FPercentPoint*

8.7   Test data for Algorithms

Algorithm 8.1, *Binomial*. Input: $n = 3$, $p = 0.4$, $x =$ (i) 0, (ii) 1, (iii) 2, (iv) 3. Output: *Binomial* = (i) 0.216, (ii) 0.432, (iii) 0.288, (iv) 0.064.

Algorithm 8.2, *SimpsonIntegral*. Input: Define a function $0.5x^2 - 5x + 14$, $a = 2$, $b = 14$. Output: *SimpsonIntegral* = 144.

Algorithm 8.3, *Normal*. Input: $z =$ (i) 1.0, (ii) -0.5. Output: *Normal* = (i) 0.2420, (ii) 0.3521.

Algorithm 8.4, *NormalProb*. Input: (i) $z = 1$; (ii) $z = 2$. Output: *NormalProb* = (i) 0.841; (ii) 0.977.

Algorithm 8.5, *NormalProbApprox*. Input: $z =$ (i) 1, (ii) 2. Output: *NormalProbApprox* = (i) 0.841, (ii) 0.977. Compare with output for test of Algorithm 8.4.

Algorithm 8.6, *NormalPercentPointApprox*. Input: $p =$ (i) 0.05; (ii) 0.95; (iii) 0.99. Output: *NormalPercentPointApprox* = (i) -1.64449 (exact -1.64485); (ii) 1.64449; (iii) 2.32765 (exact 2.32635).

Algorithm 8.7, *LnGamma*. Input: $w =$ (i) 4, (ii) 5.5. Output: *LnGamma* = (i) 1.79176, (ii) 3.95781.

Algorithm 8.8, *tProb*. Input: (i) $t = 2.776$, $k = 4$. (ii) $t = 2.528$, $k = 20$. Output: *tProb* = (i) 0.974989; (ii) 0.990000.

Algorithm 8.9, *ChiSquaredProb*. Input: (i) $x = 13.28$, $k = 4$, (ii) $x = 31.41$, $k = 20$. Output: *ChiSquaredProb* = (i) 0.990014, (ii) 0.949994.

Algorithm 8.10, *FProb*. Input: (i) $f = 18.51$, $k1 = 1$, $k2 = 2$. (ii) $f = 4.41$, $k1 = 20$, $k2 = 10$. Output: *FProb* = (i) 0.95; (ii) 0.99.

Algorithm 8.11, *FProbApprox*. Input: (i)$f = 18.51, k1 = 1, k2 = 2$. (ii)$f = 4.41, k1 = 20$, $k2 = 10$. Output: *FProbApprox* = (i) 0.952772 (exact 0.95); (ii) 0.989502 (exact 0.99).
Algorithm 8.12, *tPercentPointApprox*. Input: (i) $p = 0.95, k = 10$. (ii) $p = 0.99, k = 20$.
Output: *tPercentPointApprox* = (i) 1.812 (exact 1.812); (ii) 2.530 (exact 2.528).
Algorithm 8.13, *ChiSquaredPercentPointApprox*. Input: (i) $p = 0.95, k = 10$. (ii) $p = 0.99, k = 20$. Output: *ChiSquaredPercentPointApprox* = (i) 18.2894 (exact 18.3070); (ii) 37.6040 (exact 37.5662).
Algorithm 8.14, *FPercentPointApprox*. Input: (i) $p = 0.95, k1 = 10, k2 = 15$. (ii) $p = 0.99, k1 = 15, k2 = 10$. Output: *FPercentPointApprox* = (i) 2.54 (exact 2.54); (ii) 4.62 (exact 4.56).
Algorithm 8.15, *FPercentPoint*. Input as for Algorithm 8.14. Output is the exact value stated.

8.8   Exercises

1   Write a program, based on Algorithm 8.1, *Binomial*, to calculate the probability that a binomial random variable is less than or equal to x_0 given the values of x_0, n and p (the cumulative distribution function).

2   Write a program to calculate the probability mass function of the Poisson distribution, as described in Section 8.2.2. Run the program and generate the probabilities, up to P(3), for a distribution with parameter $\lambda = 2/3$. (The values are P(0) = 0.5134, P(1) = 0.3423, P(2) = 0.1141, P(3) = 0.0254; the total probability of higher values is 0.0048.)

3   Write a program to calculate the probability mass function of the geometric distribution, as described in Section 8.2.3. Run the program and generate the probabilities, up to P(12), for a distribution with parameter $p = 1/4$. (The values are 0.25, 0.1875, 0.1406, 0.1055, 0.0791, 0.0593, 0.0445, 0.0334, 0.025, 0.0188, 0.0141, 0.0106; the total probability of higher values is 0.0316.)

4   Use the functions defined in Section 8.3 to find the probability density function and the cumulative distribution function for the exponential distribution with parameter $\lambda = 1.2$ for $x = 0.25, 0.5, 0.75, 1.0, 1.25, 1.5, 2.0, 3.0, 5.0$.

5   Use the function defined at the end of Section 8.3 to find the median, and upper and lower quartiles, of the exponential distribution with parameter $\lambda = 0.8$.

6   The binomial probabilities arising from a distribution in which n is large and p is small can be approximated by the corresponding Poisson probabilities, with $\lambda = np$. Write a program to calculate the probabilities up to P(10), by both methods, and the differences between the values found using each of the two methods. Print the results in a convenient table to be used for comparison of the · exact and approximate probabilities.
   Run the program for (i) $n = 100$, $p = 0.01$; (ii) $n = 50$, $p = 0.04$; (iii) $n = 25$, $p = 0.05$.

7   Write a program which uses Algorithm 8.2, *SimpsonIntegral*, to find the (cumulative) distribution function of the random variable X whose probability density is $f(x)$ $= \dfrac{1}{108} x(6 - x)^2$ for $0 \le x \le 6$ and zero elsewhere. Print out values of the distribution function at intervals of 0.5 for X.

The **mean** of a continuous distribution is given by $\int x f(x)\, dx$. Write a program to find this in a similar way. (The exact value is 2.4.)

8 Write a program which uses Algorithm 8.2, *SimpsonIntegral*, to find the distribution function of the t-distribution. (Compare with Algorithm 8.4.)

9 If a variable Y follows a normal distribution with mean μ and variance σ^2 (i.e. standard deviation σ) then $Z = (Y - \mu)/\sigma$ follows the standard normal distribution, with mean 0 and variance 1 (*ABC*, Section 14.5); Z is the normal variable for which algorithms have been given in Section 8.5. Modify Algorithms 8.4 (or 8.5) and 8.6 to deal with a variable Y whose mean and variance are given as part of the input.

Run these programs to answer the following questions.
(a) If Y has mean 100 and standard deviation 15, find (i) $P(Y > 120)$, (ii) $P(Y < 125)$, (iii) $P(85 < Y < 100)$, (iv) $P(110 < Y < 140)$, (v) $P(65 < Y < 75)$, (vi) $P(Y > 90)$.
(b) If Y has mean 5 and variance 4, find the deciles of the distribution of Y (see Section 3.7).

10 Use Algorithm 8.12, *tPercentPointApprox*, to generate a t-table giving values of t corresponding to $p = 0.95, 0.975, 0.99, 0.995, 0.999$ and 0.9995, and print the table to show these in six columns; the rows of the table should correspond to the degrees of freedom. Label the table so that it could be used in the same way as a t-table printed in a book (e.g. *ABC*, Table A3).

11 Use Algorithm 8.13, *ChiSquaredPercentPointApprox*, to generate a χ^2 table giving the values of χ^2 corresponding to $p = 0.995, 0.975, 0.05, 0.025, 0.01, 0.005, 0.001$, and print the table to show these in seven columns; the rows of the table should correspond to the degrees of freedom. Label the table so that it could be used in the same way as a χ^2-table printed in a book (e.g. *ABC*, Table A4).

12 Write a program which will give approximate values of the distribution functions (i.e. probabilities) of the normal, t, χ^2 and F variables (see end of Section 8.6.4).

9 Generation of random samples from non-uniform distributions

9.1 Introduction

When carrying out stochastic simulations, we need to be able to draw random samples from a wide range of statistical distributions. For example, if we wish to simulate people arriving at random at a railway station ticket office, we might make the time interval between arrivals an exponential random variable. Likewise, if we are simulating the engineering performance of a temporary footbridge, we might assume that the weights of the people crossing it are normally distributed. We shall see that the basis of all the methods of sampling from distributions is a source of random samples from the uniform distribution $U(0, 1)$. We shall assume that these are available from a random number generator, as described in Chapter 7, or as a function on the computer. The references given at the beginning of Chapter 7 are also relevant to this chapter.

9.2 Simulation of the behaviour of a queue

Suppose we wish to discover the effect on the queue at a railway booking office of allowing customers to pay for their tickets by cheque as an alternative to paying by cash. To program a simple queueing system such as this we need to specify, or be able to calculate, the arrival time $ArriveTime_i$ and the service time $ServeTime_i$ of each customer i. Let us call the time at which the service of the ith customer is completed $DepartTime_i$. At this instant, either there is a queue, in which case $DepartTime$ for the $(i + 1)$th customer is $DepartTime_i + ServeTime_{i+1}$, or there is no queue, in which case $DepartTime$ for the $(i + 1)$th customer is $ArriveTime_{i+1} + ServeTime_{i+1}$. These calculations form the core of the simulation and are programmed in Program 9.1, *BookingOfficeSimulation*.

```
program BookingOfficeSimulation(input,output);

(* simulates the behaviour of a queue at a booking office    *)
(* with one server.  Customers arrive at random at a ·chosen *)
(* rate and pay by cash or cheque with chosen probability.   *)
(* Uses Random and RndExp.                                   *)

var ArriveTime : real; (* arrival time of customer *)
    CashTime,ChqTime : real; (* fixed times taken to serve *)
                             (* cash and cheque customers   *)
    DepartTime : real; (* time when service is completed *)
    IdleTime : real; (* total time server is not serving *)
    InterArriveTime : real; (* random time between customers *)
    MeanArriveGap : real; (* mean of inter-arrival times *)
    nCash,nChq : integer; (* number of customers paying by *)
                          (* cash and cheque *)
    nTotal : integer; (* total number of customers *)
    ProbChq : real; (* probability a customer pays by cheque *)
```

```pascal
          RunDuration : real; (* duration of simulation *)
          Seed : integer; (* seed for random number generator *)
          ServeTime : real; (* time taken to serve a customer *)
          TotalServeTime : real;
          TotalWait : real; (* total waiting time of all customers *)
          Wait : real; (* time a customer waits in queue *)


function Random(var Seed : integer) : real;

          (*** Algorithm 7.2 ***)


function RndExp(Mean:real;var Seed:integer):real;

          (*** Algorithm 9.7 ***)



begin      (*  set parameters  *)
     write('Service time - cash        ');read(CashTime);
     write('Service time - cheque      ');read(ChqTime);
     write('Probability of cheque      ');read(ProbChq);
     write('Mean inter-arrival time    ');read(MeanArriveGap);
     write('Duration of simulation     ');read(RunDuration);
     writeln('Seed for random generator ');
     write('must be positive integer  ');read(Seed);

(*   set variables to zero   *)
     nTotal := 0;
     nCash := 0;
     nChq := 0;
     ArriveTime := 0;
     DepartTime := 0;
     TotalServeTime := 0;
     TotalWait := 0;

(*   begin simulation   *)
     repeat
         nTotal := nTotal + 1;
         InterArriveTime := RndExp(MeanArriveGap,Seed);
         ArriveTime := ArriveTime + InterArriveTime;
         if Random(Seed) >= ProbChq
         then begin
             Servetime := Cashtime;
             nCash := nCash + 1
         end
         else begin
             ServeTime := ChqTime;
             nChq := nChq + 1
         end;
         if ArriveTime > DepartTime
             then DepartTime := ArriveTime + ServeTime
             else DepartTime := DepartTime + ServeTime;
         Wait := DepartTime - ArriveTime - Servetime;
         TotalWait := TotalWait + Wait;
         TotalServeTime := TotalServeTime + ServeTime;
     until ArriveTime >= RunDuration;
```

```
(*   end of simulation   *)

    if RunDuration > DepartTime then DepartTime := RunDuration;
    IdleTime := DepartTime - TotalServeTime;

    writeln('No of customers    ',nTotal:4);
    writeln('Cash customers     ',nCash:4);
    writeln('Cheque customers   ',nChq:4);
    writeln('Mean waiting time  ',TotalWait/nTotal:4:0);
    writeln('Mean serving time  ',TotalServeTime/nTotal:4:0);
    writeln('Total idle time    ',IdleTime:4:0)
end.
```

Program 9.1 *BookingOfficeSimulation*

In our model we shall assume that the time interval between successive customers arriving follows an exponential distribution, which is appropriate if they are arriving at random instants in time but at a constant overall rate. The random sample from the distribution is chosen using the function *RndExp* (Algorithm 9.7). The service time (*ServeTime*) for a customer will depend on whether he pays by cheque or by cash. We shall make the very simple assumption that the service time by each of these methods is a fixed value (the value for cheque obviously being greater than that for cash); it would, of course, in practice be more realistic to assume that each of these two service times had a distribution. The probability that any individual customer will pay by cheque is fixed at the beginning of the simulation, and the random number function is used to decide whether a particular customer pays by cheque or cash; this determines his service time.

The variables *Total Wait* and *TotalServeTime* are introduced to record interesting characteristics of each run of the simulation: *Total Wait* records the sum of the times that individual customers spend waiting in the queue, while *TotalServeTime* is the sum of the service times of the customers. The mean values, obtained by dividing the totals by the total number of customers, are output at the end of the program. The total time for which the server (i.e. the clerk in the booking office) is not actually serving is also of interest: it is output as *IdleTime*.

Let us consider a particular run of this simulation. Suppose that we wish to simulate what happens at a booking office between 8.00 and 9.00 a.m. Using seconds as units, the duration of this run is 3600. Reasonable values for service times are, for cash, 15 seconds and, for cheque, 45 seconds. Let us take the mean time-interval between arrivals as 20 seconds. Five runs of this simulation were made with no customers paying by cheque, and the mean waiting times were

22.1, 39.2, 16.7, 17.4, 11.9 seconds.

Five more runs were made, under the same conditions except that there was a probability 0.10 that any individual customer would offer a cheque. The mean waiting times now were

66.0, 95.9, 59.0, 201.4, 105.2 seconds.

The magnitude of the mean waiting time is much greater in the second set of runs than in the first, and so is the variability in waiting time.

It would be interesting to look at the distribution of the waiting times of individual customers within a run, and this information would be easy to obtain. If we wish to make a number of runs with the same parameter values, we can add an input statement at the end of the program asking whether the program should be run again. Program 9.1, with changes of detail, would make a general program for any single-server queue simulation.

9.3   A general method of sampling statistical distributions

The simplest random variable to sample by computer, given a supply of uniformly distributed random variables, is the one that takes the value 1 with probability p and the value 0 with probability $(1 - p)$. This is known as the Bernoulli distribution (*ABC*, Section 7.5); it may be regarded as a binomial distribution in which the parameter n equals 1. Bernoulli distributions might be used in a simulation of a game of tennis (*ABC*, Section 11.6). Thus we might fix the probability that one of the players wins directly from a serve at 0.1. Then, in the simulation, we sample a Bernoulli distribution with $p = 0.1$; if the value of the random variable is 1, the player chosen wins directly from his serve, but does not if the value is zero. Similarly, we might fix the probability that a particular player wins a rally and again decide the progress of a game by sampling a Bernoulli distribution. To choose the sample from the Bernoulli distribution, given a uniform variable u, we make the variable 1 if $u < p$ and 0 otherwise.

It may help the reader to understand better the process of drawing a random sample from a statistical distribution if we consider how it was sometimes done before computers were common. To draw a random sample from a Bernoulli distribution with $p = 0.1$ we might put ten discs in a hat, label one of them '1' and label the remaining nine '0'. After a thorough shake of the hat, the label of '0' or '1' on a disc chosen from the hat would give us the sample value. Let us now consider sampling a more complex distribution, a binomial distribution with $n = 3$, $p = 0.4$. The probabilities of obtaining each possible value of the variable are given in the second line of Table 9.1, and the cumulative probabilities in the third line.

We might draw the sample by putting into a (very large) hat 1000 discs, of which 216 are marked '0', 432 marked '1', 288 marked '2' and 64 marked '3'. It would not be necessary to have special marks on the discs provided that they were numbered 1–1000. If we draw out of the hat any of the discs numbered 1–216, then we use a sample value of 0; the discs 217–648 give a sample value of 1; 649–936 give 2; and 937–1000 give 3.

We can represent this method using a graph (Figure 9.1). Choosing a disc corresponds to choosing a number (413 in Figure 9.1) between 1 and 1000, and may be denoted by a point on the left-hand axis. The fact that 413 falls in the interval 217–648 inclusive, and therefore corresponds to the sample value 1, is seen on the graph by drawing a line, parallel to the x-axis, to meet the graph. Now the graph is, in fact, the graph of $F(x)$, the cumulative distribution function of X; and we see that if the random number u, when scaled to lie between 0 and 1, falls in the interval

$$F(x_{i-1}) < u \leqslant F(x_i),$$

then x_i is the appropriate sample value from the distribution. This is a general method that may be used with any discrete distribution.

Table 9.1

x	0	1	2	3
$\Pr(X = x)$	.216	.432	.288	.064
$\Pr(X \leq x)$	.216	.648	.936	1.000

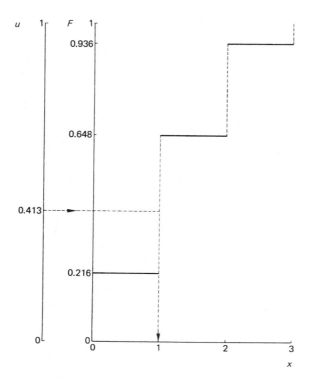

Figure 9.1

By considering a similar graph, we may develop a method of choosing a random sample from a continuous variable. Imagine a discrete random variable in which the intervals between successive values are very small. The graph of $F(x)$ then changes by steps, but very small ones, and in the limit the graph, for a continuous distribution function, will be a continuous line (Figure 9.2). As with the discrete random variable, we may proceed in a similar way to pass from the uniform random variable u (with a value of .237 in the example) to the value x (.270 in the example) of the random sample. The required sample value x comes from solving the equation

$$u = F(x),$$

where $F(x)$ is the cumulative distribution function of the random variable. The solution

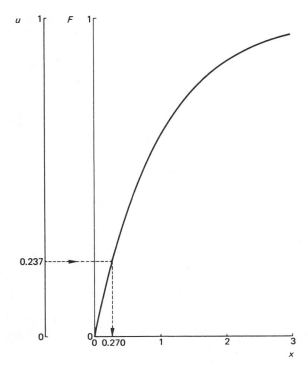

Figure 9.2

may be written in terms of the inverse distribution function as

$$x = F^{-1}(u).$$

For example, the exponential function has the cumulative distribution function (Section 8.3)

$$F(x) = 1 - \exp(-\lambda x).$$

To obtain a random value x from the distribution, given a uniform random variable u, we solve

$$u = 1 - \exp(-\lambda x),$$

to obtain

$$x = -\ln(1 - u)/\lambda.$$

Since u is $U(0, 1)$ then $(1 - u)$ is also $U(0, 1)$. We may therefore make the argument of the log function the uniform random variable we have generated. Hence a random sample from the exponential distribution, with parameter denoted by l, is obtained by the single line of programming:

$$x: = -\ln(u)/l.$$

The mean of this exponential distribution will be $1/l$ (*ABC*, page 200). The method we have given is a general one for continuous random variables provided that we can obtain an explicit expression for the inverse distribution function $F^{-1}(x)$.

We have begun with the general methods. However, methods based on the particular properties of distributions are often more convenient and efficient. We shall therefore look individually at the statistical distributions that are commonly used in simulation.

The algorithms are written as Pascal functions except that the algorithm for the normal distribution produces two values and is therefore written as a procedure. Each algorithm requires a source of $U(0, 1)$ random variables which is provided by the function *Random* (Algorithm 7.2). If your computer has a satisfactory random number generator that produces values u of a $U(0, 1)$ distribution, then – assuming the generator is also called *Random* – the only change required will be to replace '*Seed*: integer' by '*u*: real' in the argument list of each of the algorithms that we give.

9.4   Discrete random variables

9.4.1   Bernoulli distribution
This serves as a model for any situation in which we can think of a trial being made with probability p of 'success' and probability $(1-p)$ of 'failure' (*ABC*, Section 7.5). Thus we may use it to model such widely different situations as the toss of a coin ('success' being a Head), or whether a surgical operation is successful or not. The standard definition gives the variable a value of 1 with probability p and a value 0 with probability $(1-p)$, and we follow this definition in Algorithm 9.2, *RndBernoulli*.

```
function RndBernoulli(p : real;var Seed : integer) : integer;

(* generates a random sample from a Bernoulli distribution *)
(* with parameter p.  Uses function Random.              *)

begin
    if Random(Seed) < p
        then RndBernoulli := 1
        else RndBernoulli := 0
end; (* of RndBernoulli *)
```

Algorithm 9.2 *RndBernoulli*

9.4.2   Binomial distribution
The probability mass function of this variable has already been given in Section 8.2 (see also *ABC*, Section 7.6). The variable may be used to model, for example, the number of defective components in random samples of 10 taken from an industrial production line, or the number of blue-eyed players (as opposed to non-blue-eyed players) in cricket teams. The most convenient way to generate random samples is to use the property that a binomial distribution with parameters n and p is the sum of n Bernoulli variables, each with parameter p. Thus we take a $U(0, 1)$ variable u. If $0 < u < p$ count 1, otherwise count 0. Repeat n times, and the sum of the counts is a binomial variable. This is programmed in Algorithm 9.3, *RndBinomial*.

```
function RndBinomial(NoOfTrials : integer;p : real;
                     var Seed : integer) : integer;

(* generates a random sample from a binomial distribution *)
(* with parameters NoOfTrials and p. Uses function Random.*)

var Count : integer;
    i : integer; (* index of loop *)

begin
    Count := 0;
    for i := 1 to NoOfTrials do
        if Random(Seed) < p then
        Count := Count + 1;
    RndBinomial := Count
end; (* of RndBinomial *)
```

Algorithm 9.3 *RndBinomial*

9.4.3 Poisson distribution

This may be used, for example, to model the number of misprints on the page of a book, or the number of cars that pass an observation point on a road in one minute. It is described in *ABC*, Chapter 19; the probability mass function has already been given on page 74 of this book.

We use the fact that if events occur at random in time, then the interval between events has an exponential distribution and the number of events in a fixed interval of time has a Poisson distribution. If we have a source of random variables from an exponential distribution, which are easy to generate, and we find how many just fit into a unit time interval, then that number will have a Poisson distribution. More precisely, if an exponential variable is labelled v_i, and if we fix X such that

$$\sum_{i=0}^{X-1} v_i < 1 \le \sum_{i=0}^{X} v_i \qquad (X = 0, 1, \ldots)$$

then X has a Poisson distribution. This Poisson distribution will have parameter λ (and therefore mean λ) if we sample from an exponential distribution with parameter λ, and therefore mean $1/\lambda$.

To obtain a random variable v_i from an exponential distribution with parameter λ, using a uniform random variable u_i, we put (see page 97)

$$v_i = -(\ln u_i)/\lambda.$$

Thus the Poisson variable X we require is given by

$$\sum_{i=1}^{X} -(\ln u_i)/\lambda < 1 \le \sum_{i=1}^{X+1} -(\ln u_i)/\lambda$$

or $\quad \ln \prod_{i=0}^{X-1} u_i > -\lambda \ge \ln \prod_{i=0}^{X} u_i$

or $\quad \prod_{i=0}^{X-1} u_i > e^{-\lambda} \ge \prod_{i=0}^{X} u_i.$

We take a uniform variable u_0; if it is less than or equal to $e^{-\lambda}$ we stop, and take $X = 0$. If $u_0 > e^{-\lambda}$, we form the product u_0u_1; if this is less than or equal to $e^{-\lambda}$, we take $X = 1$. Otherwise we continue until $u_0u_1u_2 \ldots u_x$ is the first product that is less than or equal to $e^{-\lambda}$, which leads us to take $X = x$. This method is programmed in Algorithm 9.4, *RndPoisson*.

```
function RndPoisson(Mean : real;var Seed : integer) : integer;

(* generates a random sample from a Poisson distribution *)
(* with stated mean.  Uses function Random.            *)

var Count : integer;
    ProbZero : real;
    Product : real;

begin
    Count := 0;
    Product := Random(Seed);
    ProbZero := exp(-Mean);
    while Product > ProbZero do
    begin
        Count := Count + 1;
        Product := Product*Random(Seed)
    end;
    RndPoisson := Count
end; (* of RndPoisson *)
```

Algorithm 9.4 *RndPoisson*

9.4.4 Geometric distribution

If independent Bernoulli trials are made until a 'success' occurs, then the total number of trials required is a geometric random variable. The distribution is described in *ABC*, Section 7.7, and its probability mass function has already been given on page 74 of this book. It could serve, for example, as a model for the number of cars passing an observation point up to the first car of a particular make: record this number and begin counting again. Algorithm 9.5, *RndGeometric*, is based directly on this definition. If p is small this algorithm will be slow; Knuth (1981) suggests an alternative. Algorithm 9.5

```
function RndGeometric(p : real;var Seed : integer) : integer;

(* generates a random sample from a geometric distribution *)
(* with parameter p.  Uses function Random.              *)

var Count : integer;

begin
    Count := 1;
    while Random(Seed) > p do
        Count := Count + 1;
    RndGeometric := Count
end; (* of RndGeometric *)
```

Algorithm 9.5 *RndGeometric*

is replaced by an algorithm whose operative statements are

$$u: = Random\,(Seed);$$
$$RndGeometric: = \text{trunc}\,(\ln{(u)}/\ln{(1-p)})+1.$$

This is based on the fact that the integer part, plus one, of an exponential distribution is a geometric distribution.

9.5 Normal distribution

The normal distribution is the most important distribution in statistics. It is discussed in *ABC*, Chapter 14; the probability density function has already been given on page 77 of this book. It is likely to provide a good model for a variate when:

1) there is a strong tendency for the variate to take a central value;
2) positive and negative deviations from this central value are equally likely;
3) the frequency of deviations falls off rapidly as the deviations become larger.

It also provides a good approximation to many other statistical distributions over certain ranges of their parameters. Examples of variates which it might model are the weights of packets of cornflakes and the heights of members of a football team.

We confine ourselves to obtaining random samples from the standard normal distribution, i.e. the distribution with mean 0 and variance 1 whose values are often represented by z. If we require random samples from the normal variable x which has mean μ and variance σ^2, we derive x from z by the relation

$$x = \mu + z\sigma.$$

9.5.1 Use of the central limit theorem
The simplest method of generating normal variables is an approximate one, and relies on the central limit theorem (*ABC*, Section 15.1). A U(0, 1) variable, which we denote by u, has mean 0.5 and variance $\frac{1}{12}$ (*ABC*, Section 13.5). Hence $\sum_{i=1}^{n} u_i$, the sum of n independent uniform variables, has mean $n/2$ and variance $n/12$ (*ABC*, Section 15.2). Moreover, by the central limit theorem, this sum is approximately normally distributed provided that n is sufficiently large; in this particular case, the value 12 of n is adequate and has the convenience of keeping the arithmetic simple. Thus the sum of 12 independent uniform variables is approximately normal with mean 6 and variance 1. Subtracting 6 from this sum gives a standard normal variable. The method does not produce values outside the range $(-6, 6)$, but this is not likely to matter since such values have an extremely low probability of occurring in the standard normal distribution.

9.5.2 The polar method
Box and Muller proposed the following method, which is based on the fact that if u_1, u_2 are independent samples from U(0, 1), then

$$z_1 = (-2\ln u_1)^{1/2}\sin 2\pi u_2,$$
$$z_2 = (-2\ln u_1)^{1/2}\cos 2\pi u_2$$

are independent samples from $N(0, 1)$. (A proof is given in Fishman, 1978.) Marsaglia and Bray modified the Box-Muller method to avoid the use of trigonometric functions. Their method is:

(1) generate two $U(0, 1)$ random variables u_1, u_2;
(2) calculate $w_1 = 2u_1 - 1$, $w_2 = 2u_2 - 1$;
(3) calculate $w = w_1^2 + w_2^2$; if $w \geq 1$ return to (1), otherwise continue;
(4) calculate $c = (-2 \ln w/w)^{1/2}$, $z_1 = c * w_1$, $z_2 = c * w_2$.

The first three steps choose a random point in the circle with centre the origin and unit radius. The point has Cartesian coordinates (w_1, w_2) and, in the usual notation, polar coordinates (r, θ). It may be shown that r^2 $(= w)$ and $\theta/2\pi$ are independent $U(0, 1)$ variables. If, in the Box-Muller formula, we substitute $u_1 = w$, $u_2 = \theta/2\pi$ (and use the relations $\sin 2\pi u_2 = \sin \theta = w_1/\sqrt{w}$; $\cos 2\pi u_2 = w_2/\sqrt{w}$) we find that z_1, z_2 are as described in Step (4) above. This method is called the polar method. The method is programmed in Algorithm 9.6, *GenerateNormal*.

```
procedure GenerateNormal(var RndNormal1,RndNormal2 : real;
                         var Seed : integer);

(* generates two random samples from a standard Normal *)
(* distribution.  Uses function Random.              *)

var c,w,w1,w2 : real; (* variables used in evaluation *)

begin
    repeat
        w1 := 2*Random(Seed) - 1;
        w2 := 2*Random(Seed) - 1;
        w  := w1*w1 + w2*w2
    until w < 1;
    c := sqrt(-2*ln(w)/w);
    RndNormal1 := c*w1;
    RndNormal2 := c*w2
end; (* of GenerateNormal *)
```
Algorithm 9.6 *GenerateNormal*

The method is exact only if the two uniform variables are independent; if the random numbers are generated by a linear congruential generator, for example, this will not be strictly true. We give the polar rather than the Box-Muller method because it is less susceptible to this failing. The safest way to resolve the difficulty is to use two independent random number generators and take u_1 from one and u_2 from the other. Another approach, using a single random number generator, is not to take successive values from the generator but to allow a gap of up to (say) five, the exact size chosen at random, between one number and the next.

9.6 Distributions based on the normal distribution

Random variables from the sampling distributions χ^2, t and F may easily be generated using the definition of these variables in terms of normal variables. Using z_i to denote a standard normal variable then, provided that all the z_i's are independent,

1) the chi-squared variable with k degrees of freedom is given by

$$\chi^2 = z_1^2 + z_2^2 + \ldots + z_k^2;$$

2) the t variable with k degrees of freedom is given by

$$t = z_0 \sqrt{k/(z_1^2 + z_2^2 + \ldots + z_k^2)};$$

3) the F distribution with k_1 and k_2 degrees of freedom is given by

$$F = \frac{\chi_{k_1}^2/k_1}{\chi_{k_2}^2/k_2}.$$

The **lognormal distribution** is sometimes useful in simulations. It has been used, for example, as a model for the distribution of (a) personal incomes, (b) age at first marriage, (c) tolerance to poison in animals. The lognormal variable Y is defined as a variable such that $\ln Y$ has a normal distribution. Thus if x is a sample from a normal distribution, then

$$y = e^x$$

is a sample from a lognormal variable.

9.7 Other continuous distributions

9.7.1 Exponential distribution
The generation of random samples from the exponential distribution was discussed in Section 7.3. The method is repeated here for reference. The distribution is described in *ABC*, Section 13.6; the probability density function is given on page 75 of this book. The distribution is frequently used to model the time interval between successive random events. Given a sample u from a uniform variable, a sample from an exponential distribution with parameter λ is given by

$$x = -(\ln u)/\lambda.$$

Since the mean, μ, of the exponential distribution is equal to $1/\lambda$, we can also write the random sample in the form $x = -\mu \ln u$, and this will usually be the most convenient form (*RndExp*, Algorithm 9.7).

```
function RndExp(Mean : real;var Seed : integer) : real;

(* generates a random sample from the exponential *)
(* distribution with stated mean.  Uses Random.   *)

var u : real;

begin
    u := Random(Seed);
    RndExp := -Mean*ln(u)
end; (* of RndExp *)
```

Algorithm 9.7 *RndExp*

9.7.2   Gamma distribution

The density function of the exponential distribution has a mode at zero. When modelling the distribution of the life-times of a product such as an electric light bulb (i.e. the length of time the bulb lasts before failing), or the serving-time taken at a ticket office or a shop counter, a distribution having a mode at a greater value than zero is useful; the gamma distribution is one such. The probability density function $G(\alpha, \beta)$ was given on page 81. We restrict ourselves to gamma distributions having an integer value for the 'shape' parameter α.

We use the properties that the sum of k independent variables, each from a $G(\alpha, \beta)$ distribution, is a $G(k\alpha, \beta)$ distribution (Hogg and Craig, 1970); and that $G(1, \beta)$ is an exponential distribution with parameter $1/\beta$ (that is with mean β). Thus, denoting uniform random variables by u_i,

$$x = -\beta \sum_{i=1}^{\alpha} \ln u_i = -\beta \ln \prod_{i=1}^{\alpha} u_i$$

is a sample from $G(\alpha, \beta)$.

Note that a χ^2 distribution with k degrees of freedom is $G\left(\dfrac{k}{2}, 2\right)$.

[*Note*:   We give no test data for the algorithms of this chapter, but some of the exercises below will serve this purpose.]

9.8   Exercises

1   Write a program to simulate the progress of a game of tennis (*ABC*, Section 11.6) between two players, A and B. When player A serves, he has probability P1 of winning the point directly; similarly player B has probability P2 of winning directly from his service. If a rally takes place, A has probability P3 of winning it. This section of the program determines who wins each single point; run it several times, and print out the winner each time. Examine the effect of changing P1, P2 and P3.

Using the standard system of tennis scoring, the next section of the program should simulate a complete game, printing out the score after each point and recording the winner; all services in one game are made by the same player. For the next game, the other player serves. Build up in this way the progress of a set, the winner being the player who reaches 6 games first *unless* the other player has by then won 5; in this case the set goes to 7–5 or 7–6 before it is decided.

Run the complete program several times, varying P1, P2 and P3, and printing suitable descriptive output.

2   Modify Program 9.1, *BookingOfficeSimulation*, to allow for a third category of customer, who pays by credit card; allow 60 seconds for this transaction.

3   Add to Program 9.1 instructions to print out the time, and the length of the queue, as each customer arrives.

4   Modify Program 9.1 so that a second service-point is opened when the queue-length exceeds some prescribed value Q; the two service-points continue operating until one has no customers, and it then closes.

5 Modify Program 7.1, *CardCollection*, to include an element of 'rarity', for example collecting a set of stamps of a foreign country when the smaller-priced stamps are easier to find than the larger-priced ones.

6 Rewrite Program 9.1, *BookingOfficeSimulation*, to allow the service times to have normal distributions with the means 15 s and 45 s previously used; set the standard deviations equal to 3 s and 7 s respectively.

Extend question 2 in a similar way, making the credit card transaction time normally distributed with mean 60 and standard deviation 10 s. [Refer to Section 9.5 on normal variables with general values for mean and variance.]

7 Write a program which:
 i) generates a random sample from a named distribution (using a function);
 ii) generates k such random samples and finds their mean;
 iii) repeats this n times;
 iv) forms a frequency table of the resulting n means;
 v) plots a histogram of these means.

8 Run the program described in question 7 with a uniform distribution on the interval (0, 1) as the named distribution. Run simulations of $n = 100$ samples with k taking the values (i) 2, (ii) 6, (iii) 12. Do the three frequency tables and histograms produced resemble those of normal distributions? [See Exercises 10.7, question 10, for a goodness-of-fit test for a normal distribution.]

9 Repeat question 8 with an exponential distribution whose mean is 1 as the named distribution.

10 Run Algorithm 9.4, *RndPoisson*, (i) with mean $\lambda = 0.75$, (ii) with $\lambda = 4$. Repeat each several times, build up frequency distributions, and compare them with the Poisson distributions with the same values of λ (see Section 8.2.2).

11 Run Algorithm 9.5, *RndGeometric*, (i) with $p = 0.25$, (ii) with $p = 0.6$. Repeat each several times, build up frequency distributions, and compare them with the geometric distributions with the same values of p (see Section 8.2.3).

12 *Unbiased estimation of variance.* Generate three random samples from a normal distribution with mean 5 and variance 9. Find the mean of these three, and also their variance (i) using one of the methods in Chapter 6 with $(n-1)$, i.e. 2, as divisor; (ii) replacing $(n-1)$ by n, i.e. 3. Repeat this 100 times, and build up a frequency distribution for the mean of the three, and for the variance calculated by both methods. Calculate the mean and variance in each of these three frequency distributions.

13 Program Box and Muller's method of generating normal variables (Section 9.5.2). In order to lessen the effect of non-independence when using congruential generators, modify the program so that the values of u_1 and u_2 are not successive values from the random number generator, but are values separated by a gap of random length from one to five. Run the two programs several times and compare results.

14   Write a program to draw random samples from a gamma distribution, as described in Section 9.7.2. Run it with (i) $\alpha = 5$, $\beta = 3$; (ii) $\alpha = 6$, $\beta = 7$. Also use it to draw random samples from a χ^2 distribution with 6 degrees of freedom.

   In each case take a large number of samples and summarise the results in a grouped frequency table.

15   Draw a flow diagram for Program 9.1, *BookingOfficeSimulation*, including the modification described in question 2.

   Make any changes needed to include the random variables specified in question 6.

10 Significance tests and confidence intervals

10.1 Introduction

We now consider two methods which the statistician may use in order to make inferences about a population based on observations from a sample that he has collected: significance testing (*ABC*, Chapters 12, 16, 18), and estimation by calculating confidence intervals (*ABC*, Chapter 17). As a basis for a significance test, we set up a null hypothesis and make suitable assumptions about our data; then we obtain a suitable statistic, such as a t-statistic, or a χ^2-statistic, whose sampling distribution is known provided that our assumptions are justified. The value of the statistic is then calculated for our particular set of data, and this is compared with tables of the theoretical sampling distribution to complete the test. After we have calculated obvious summary statistics of a set of data, such as the mean and variance, determining the value of a test statistic (such as t) is often a trivial calculation for a computer, and will occupy only a few lines of program.

But the actual outcome of a significance test is just one item of evidence in interpreting a set of data. We must be sure to output, as well as the value of the test statistic, all the relevant information we need to set the significance test in context. The computer can be made to print out critical values or p-values of the sampling distributions of the statistics. We have not written these procedures into our algorithms for test statistics, but we shall indicate how the algorithms that we gave in Chapter 8 for probability functions may be used to produce this information.

When finding a confidence interval for a population parameter, such as a mean, we need an estimate of the parameter, such as the sample mean, and we also need to know what is the sampling distribution of this estimator. In many commonly useful cases, provided that certain standard assumptions hold, the sampling distribution will be a normal or a t distribution, whose mean is the unknown population parameter. When this is so, the confidence interval is given by the estimate plus or minus t standard errors, where t is a factor that is usually found from tables. By using a computer, it is possible to avoid looking up tables, and calculate the factor t to an adequate level of accuracy.

We consider algorithms according to the type of data available: single sample, two samples (paired or unpaired), or frequency data. Since the basic arithmetic for significance testing and for confidence intervals is similar, it is often convenient to consider both processes together for computing purposes, even though they are using the sample data in different ways.

10.2 Single sample analysis

Suppose we have an array of data containing n observations of a random variable X which together form a random sample from a population. If in this population X is normally distributed, or approximately so, we may use a *t*-statistic to test whether the observed mean $\bar{x}$ differs significantly from a specified value μ (*ABC*, Section 16.11), and we may also construct a confidence interval for the true mean μ (*ABC*, Section 17.5). We give algorithms for both these procedures.

10.2.1   The *t*-test

When a random sample of *n* items is available from a population, the sample mean being $\bar{x}$ and estimated variance s^2, we may wish to test the null hypothesis that the true population mean is μ_0. This requires the *t*-statistic

$$t = \frac{(\bar{x} - \mu_0)}{\sqrt{s^2/n}},$$

whose sampling distribution is *t* with $(n-1)$ degrees of freedom. Algorithm 10.1, *tTestOnMean*, prints out the statistic and other relevant information, given the four quantities on the right-hand side of the above formula.

```
procedure tTestOnMean(SampleMean,EstVariance,NullMean : real;
                      n : Units );

(* prints the relevant information for making a t test when *)
(* comparing a sample mean of n observations with the mean  *)
(* on a null hypothesis given sample estimate of the        *)
(* variance of the observations.                            *)

const Dp = 2;
      Field = 8;

var StdError : real;
    tStatistic :real;

begin
    StdError := sqrt(EstVariance/n);
    tStatistic := (SampleMean-NullMean)/StdError;
    writeln;
    writeln('t-test on mean');
    writeln;
    writeln('Mean on null hypothesis    ',NullMean:Field:Dp);
    writeln;
    writeln('Sample mean                ',SampleMean:Field:Dp);
    writeln;
    writeln('Standard error of mean     ',StdError:Field:Dp);
    writeln;
    writeln('t-statistic                ',tStatistic:Field:Dp);
    writeln;
    writeln('Degrees of freedom         ',n-1:Field)
end; (* of tTestOnMean *)
```

Algorithm 10.1 *tTestOnMean*

Information on the sampling distribution may be obtained by using the algorithms for the *t* distribution from Chapter 8. The critical value of *t* needed in a 5% two-tail significance test can be found by inputting $p = .975$ into *tPercentPointApprox* (Algorithm 8.12). In general for a two-tail test the probability value *p* which is input is such that the upper tail probability $(1 - p)$ equals half the significance level; for a one-tail test $(1 - p)$ must be set equal to the significance level. Alternatively, we may wish to obtain the probability, or p-value, that our particular calculated t-value for a sample is exceeded in absolute value. To do this, we input the absolute value of our t statistic into *tProb* (Algorithm 8.8), and obtain as output p. Then $1 - p$ is the appropriate probability for a one-tail test, and $2(1 - p)$ for a two-tail test.

10.2.2 Confidence interval for mean

A 95% central confidence interval for the true population mean μ is given by the expression (*ABC*, Section 17.5)

$$\bar{x} - \sqrt{s^2/n} \cdot t_{(n-1, 0.05)} \leq \mu \leq \bar{x} + \sqrt{s^2/n} \cdot t_{(n-1, 0.05)}.$$

In this expression t stands for the value which is exceeded by the absolute value of a t variable, with $(n-1)$ degrees of freedom, with probability 0.05. Thus both tails of the t variable are taken into account. If $F(t)$ is the cumulative distribution function of the t variable with $(n-1)$ degrees of freedom, then the value of t we require is the solution of $F(t) = 0.975$. Algorithm 10.2, *ConfIntForMean*, calculates the confidence interval for any confidence level, not just for 95%; it uses *tPercentPointApprox* (Algorithm 8.12) as a function to calculate the appropriate value of t to put in the expression. One could avoid using this function by arranging to insert the appropriate t-value as part of the input.

```
procedure ConfIntForMean(SampleMean,EstVariance : real ;
                         n : Units ;
                         ConfProb : real);

(* prints the confidence interval of stated probability for *)
(* the population mean, given the mean and variance of a    *)
(* sample of n observations.                                *)
(* Uses tPercentPointApprox and NormalPercentPointApprox.   *)

const Dp = 2;
      Field = 8;

var IntMin,IntMax : real; (* endpoints of interval *)
    p : real; (* probability used to determine value of t *)
    StdError : real; (* of sample mean *)
    t : real; (* t-value in confidence interval formula *)

begin
    p := 0.5*(1.0 + ConfProb);
    t := tPercentPoint(p,n-1);
    StdError := sqrt(EstVariance/n);
    IntMin := SampleMean - t*StdError;
    IntMax := SampleMean + t*StdError;
    writeln(100*ConfProb:Field:Dp,
                        '% Confidence Interval for Mean');
    writeln('( ',IntMin:Field:Dp,' , ',IntMax:Field:Dp,' )')
end; (* of ConfIntForMean *)
```

Algorithm 10.2 *ConfIntForMean*

If the variance in the population is known, and need not be estimated from the sample (a situation common in textbooks but rare in practice!) the appropriate value from the standard normal distribution replaces *t* in the expression for the confidence interval (e.g. the value 1.96 if we require a 95 % central confidence interval).

10.3   Paired samples

For this case, we assume that two arrays *x*, *y* are given, such that, for all *i*, *x*[*i*] is paired with *y*[*i*]. There are *n* units. This is the situation when two observations, denoted by X and Y, have been recorded on each unit in the sample of *n*. Significance tests (*ABC*, Section 16.12) and confidence intervals are based on the *differences* between the paired values; these differences are assumed to be normally distributed.

To test the null hypothesis that the true means in the two populations X and Y are equal, we begin by forming the array of differences whose typical element is $d[i] = x[i] - y[i]$. We then proceed as for the single-sample case, with *d* replacing *x* and *NullMean* set equal to zero. Algorithm 6.1 or 6.2 may be used to find the mean and estimated variance of the differences in the array. Algorithm 10.1 may be used to give the *t*-statistic for the test on the mean *difference*, and similarly Algorithm 10.2 will give a confidence interval for the mean *difference*. Slight changes in the printed output would of course be needed in doing this.

10.3.1   Wilcoxon signed-rank test
This is a non-parametric test of the null hypothesis that the differences between the pairs of observations have a median of zero (*ABC*, Section 12.6). In the first part of Algorithm 10.3, *FindSignedRankStatistic*, the absolute values of the differences are

```
procedure FindSignedRankStatistic(x,y : DataVector; n : Units;
                        var TMinus,NormalDeviate : real);

(* finds the Wilcoxon signed rank statistic TMinus given      *)
(* paired samples each of size n, in arrays x and y.   Norma  *)
(* deviate of TMinus is calculated but valid only if n>16.    *)
(* Uses FindRanks.                                            *)

const MaxReal = 1E38;
var AbsDiff : DataVector; (* absolute values of differences   *)
    Diff : DataVector; (* differences of paired observations *)
    Difference : real; (* difference of pair of observations *)
    i : integer; (* label for sample values *)
    MinusCount : integer; (* number of negative differences *)
    NonZeroCount : integer; (* number of nonzero differences *)
    Rank : DataVector; (* ranks of absolute differences *)
    TMean : real; (* mean value of TMinus on null hypothesis *)
    TVar : real; (* sampling variance of TMinus *)
    ZeroCount : integer; (* number of zero differences *)

begin
    ZeroCount := 0;
    for i := 1 to n do
```

```
begin
    Difference := x[i] - y[i];
    if Difference = 0.0
    then begin
        ZeroCount := ZeroCount + 1;
        AbsDiff[i] := MaxReal
    end
    else
        AbsDiff[i] := abs(Difference);
    Diff[i] := Difference
end;
FindRanks(AbsDiff,n,Rank);
TMinus := 0.0;
MinusCount := 0;
for i := 1 to n do
    if Diff[i] < 0.0
    then begin
        MinusCount := MinusCount + 1;
        TMinus := TMinus + Rank[i]
    end;
NonZeroCount := n - ZeroCount;
TMean := NonZeroCount*(NonZeroCount + 1)/4;
TVar := TMean*(2*NonZeroCount + 1)/6;
NormalDeviate := (TMinus - TMean)/sqrt(TVar)
end; (* of FindSignedRankStatistic *)
```

Algorithm 10.3 *FindSignedRankStatistic*

found and these are ranked. Tied observations are given average ranks. If any differences are zero, these are not used when calculating the test statistic; this can lead to awkwardness if we try to change the value of n when entering the ranking algorithm and later reset it in case it is needed elsewhere in the program. We may avoid this difficulty by replacing zero differences in the absolute difference vector by very large numbers. This ensures that the ranks of all the other differences will be the same as they would have been if the zero differences had been discarded. The number of zero differences is counted by *ZeroCount*, and the number of non-zero differences is then given by $NonZeroCount = n - ZeroCount$.

The test statistic *TMinus* is the sum of the ranks of the elements of the absolute difference vector which correspond to *negative* differences. The sampling distribution of *TMinus* is symmetric about a mean of $NonZeroCount * (NonZeroCount + 1)/4$; tables usually quote only the lower tail of the distribution of *TMinus*. If the value of *TMinus* exceeds its mean, we have an upper tail value, and in order to obtain the corresponding value in the lower tail we must subtract it from $NonZero Count * (NonZeroCount + 1)/2$; in fact this lower tail value is the sum of the ranks of the positive differences. The form of output that we recommend includes both tail values. For large values of *NonZeroCount* the sampling distribution of *TMinus* is approximately normal, and in such cases a standard normal deviate may be used for the test; this statistic is calculated in the last few lines of the algorithm.

A possible form of output follows, in which the expressions in italics show where their numerical values should appear. *TPlus* has been written instead of the actual expression: $NonZeroCount * (NonZeroCount + 1)/2 - TMinus$.

Differences	Number	Sum of ranks
Sample $1 < 2$	*MinusCount*	*TMinus*
Sample $2 < 1$	*NonZeroCount* − *MinusCount*	*TPlus*
Sample $1 = 2$	*ZeroCount*	

Standard normal deviate, *NormalDeviate*
(Valid if number of non-zero differences > 19)

10.4   Two unpaired samples

The sample of n_1 observations from one population may be stored in an array x, and the sample of n_2 observations from the other population in an array y.

10.4.1   The *t*-test

The commonly used parametric test is a *t*-test, valid when we are sampling from two normal populations. The *t*-statistic for testing the null hypothesis that the true means of the two populations are equal is (*ABC*, Section 16.13)

$$t = \frac{\bar{x}_1 - \bar{x}_2}{\sqrt{s^2 \left(\frac{1}{n_1} + \frac{1}{n_2} \right)}}$$

in which $\bar{x}_1, \bar{x}_2$ are the means of the two samples and s^2 is a pooled estimate of variance: the *t*-test is not valid (at least for small n_1, n_2) if the variances in the two populations are different. Therefore we should print out the two estimated variances s_1^2, s_2^2 for the two samples, to allow a check that they are not greatly different and that pooling the estimates is legitimate. The pooled estimate of variance is

$$s^2 = \frac{(n_1 - 1)s_1^2 + (n_2 - 1)s_2^2}{n_1 + n_2 - 2}.$$

The sampling distribution of t as defined above is a t distribution with $(n_1 + n_2 - 2)$ degrees of freedom.

   An algorithm similar to *tTestOnMean* (Algorithm 10.1) may be written. The relevant information to be output is (1) title: '*t*-test that means of two samples are equal'; (2) the means of the two samples: $\bar{x}_1, \bar{x}_2$; (3) the estimated variances of the two samples: s_1^2, s_2^2; (4) the standard error of the difference between the two means: sqrt($s^2 * (n_1 + n_2)/n_1 n_2$); (5) the *t*-statistic: t; (6) the degrees of freedom: $(n_1 + n_2 - 2)$.

10.4.2   Confidence interval for difference of means

A confidence interval for the difference between the true means of the two normal populations, in the conditions described in Section 10.4.1, is constructed in the same way as we proceeded in 10.2.2. The difference $(\bar{x}_1 - \bar{x}_2)$ replaces $\bar{x}$; s^2 is calculated from the formula in Section 10.4.1 and then s^2/n in the equation of Section 10.2.2 is replaced by $s^2 \left(\frac{1}{n_1} + \frac{1}{n_2} \right)$; and the appropriate degrees of freedom are $(n_1 + n_2 - 2)$.

10.4.3   Wilcoxon Rank-Sum (Mann-Whitney *U*) test

This is a nonparametric test of the identity of two populations from which independent random samples have been drawn (*ABC*, Section 12.5). It can therefore be applied when the normality condition in Section 10.4.1 breaks down. In Algorithm 10.4, *FindRankSumStatistic*, all the data are merged, and the rank of each observation is found using a procedure. We have suggested using *FindRanks* (Algorithm 3.5), but any other ranking procedure might be used provided that variables are set appropriately before the procedure is entered. The sum of the ranks of the observations, commonly denoted by *T*, in the first sample is found; the significance test is usually based on the statistic *U* which is equal to $T - n_1(n_1 + 1)/2$. In fact, *U* is the number of times members of sample 2 precede members of sample 1 in the joint ranking. The sampling distribution of *U* is symmetric, with mean $n_1 n_2/2$; but only the lower half of the distribution is given in tables. Thus for a particular set of data we may need the statistic $(n_1 n_2 - U)$ for comparison with the values in tables, and so we suggest outputting the *U*-statistic for each tail: these two values may be regarded as corresponding to the two samples. For large samples of data—in fact, when n_1 or n_2 is greater than 19—a normal approximation to the distribution of the *U*-statistic can be used, and the standard normal deviate for this is calculated in the final section of Algorithm 10.4.

When observations are tied in rank, an average rank is given to each observation affected: *FindRanks* (Algorithm 3.5) does this automatically. There is a correction to the

```
procedure FindRankSumStatistic(x,y:DataVector; n1,n2:Units;
                               var U,NormalDeviate : real;
                               var RankSum : real);

(* given two samples of size n1,n2 in arrays x,y, finds the *)
(* Wilcoxon rank-sum statistic using the ranked x values    *)
(* and the Mann-Whitney U statistic. U is also expressed as  *)
(* a normal deviate. This value is valid only if n1 or n2    *)
(* is greater than 19. MaxSampleSize must not be less than   *)
(* n1 + n2.   Uses FindRanks.                                *)

var i : integer; (* loop counter *)
    n : integer; (* total number of units in x and y *)
    Rank : DataVector; (* ranks of combined data *)
    UMean : real; (* mean of U on the null hypothesis *)
    UVariance : real; (* sampling variance of U *)

begin
    for i := 1 to n2 do
        x[i+n1] := y[i];
    n := n1 + n2;
    FindRanks(x,n,Rank);
    RankSum := 0.0;
    for i := 1 to n1 do
        RankSum := RankSum +Rank[i];
    U := RankSum - n1*(n1+1)/2;
    UMean := n1*n2/2;
    UVariance := n1*n2*(n+1)/12;
    NormalDeviate := (U - UMean)/sqrt(UVariance)
end; (* of FindRankSumStatistic *)
```

Algorithm 10.4 *FindRankSumStatistic*

variance used in the normal approximation which takes account of ties; its effect, however, is negligible unless the number of ties is large and so we have not incorporated it. An example of the type of output we would recommend is as follows; expressions in italics show where their numerical values should appear.

	Sample 1	Sample 2
Sum of ranks	*RankSum*	$n*(n+1)/2 - RankSum$
Sample size	*n1*	*n2*
U-statistic	*U*	$n1*n2 - U$

Standard normal deviate *NormalDeviate*
(Valid if n1 or n2 > 19)

10.5   Frequency data

There are two main types of test required with frequency data, both of them based on the chi-squared distribution.

10.5.1   Chi-squared goodness-of-fit tests

We sometimes wish to test whether a set of data could reasonably have arisen as a sample from a particular distribution. One way to do this is to group the observed data into classes, find the *observed* frequency in each class, and find also the frequency *expected* in each class if the data really do follow the particular distribution we propose for them. We compare the corresponding observed and expected frequencies using a statistic X^2 that follows a χ^2 distribution (*ABC*, Section 18.5).

```
procedure FindX2ofFit (Obs : FreqList; Exp : FreqVector;
                       NoOfClasses:Classes;
                       var X2 : real;
                       var SmallExpectation : boolean;
                       var SumDiscrepancy : boolean;
                       var X2Component : FreqVector);

(* calculates chi-squared goodness of fit statistic X2 given *)
(* observed and expected frequencies and the number of       *)
(* classes. If any expected value is found to be less than a *)
(* stated minimum the variable SmallExpectation is set to    *)
(* true. If the sum of expected frequencies does not equal   *)
(* the sum of observed frequencies then the variable         *)
(* SumDiscrepancy is set to true.                            *)

const MinimumExpValue = 5;

var Diff : real; (* observed value minus expected value *)
    ExpectedValue : real;
    i : integer; (* label of classes *)
    SumExp : real ; (* sums expected values *)
    SumObs : real ; (* sums observed values *)
```

```
begin
    SmallExpectation := false;
    SumObs := 0.0;
    SumExp := 0.0;
    for i := 1 to NoOfClasses do
    begin
        SumObs := SumObs + Obs[i];
        SumExp := SumExp + Exp[i]
    end;
    if abs(SumObs-SumExp)>0.05
        then SumDiscrepancy := true
        else SumDiscrepancy := false;
    X2 := 0.0;
    for i := 1 to NoOfClasses do
    begin
        ExpectedValue := Exp[i];
        if ExpectedValue < MinimumExpValue
            then SmallExpectation := true;
        Diff := Obs[i] - ExpectedValue;
        X2Component[i] :=Diff*Diff/ExpectedValue;
        X2 := X2 + X2Component[i]
    end
end; (* of FindX2ofFit *)
```

Algorithm 10.5 *FindX2ofFit*

In Algorithm 10.5, *FindX2ofFit*, the observed frequencies in the classes are input in an array *Obs*. The expected frequencies will already have been calculated using the appropriate probability function, and placed in an array *Exp*. The totals of observed and expected frequencies should be the same; this is tested in the algorithm, but since the sums may not be exactly equal due to rounding errors in calculating the expected frequencies, we test whether the difference *SumObs − SumExp* is small rather than exactly zero. A boolean variable, *SumDiscrepancy*, is set equal to 'true' if there is a discrepancy between the totals. The value of this variable should be checked by the calling program. Some—usually all—of the parameters of the theoretical distribution are made to equal their estimates calculated from the sample data. Thus when the theoretical distribution is the Poisson, its parameter λ, which equals the mean, will often be made equal to the mean of the sample data; similarly if the theoretical distribution is the normal, its parameters μ and σ^2, which equal the mean and variance respectively, will often be made equal to the mean and estimated variance from the sample. If we set up these correspondences for parameters, we affect the degrees of freedom of the sampling distribution of the χ^2-statistic, and hence affect the interpretation of the results. The general rule is that the number of degrees of freedom is equal to the 'number of classes, minus one for the total, minus one for each parameter estimated' (*ABC*, Section 18.6.1). This information is not used by the algorithm but the number of degrees of freedom should be output with the value of the X^2-statistic.

The X^2-statistic equals $\Sigma(Obs[i] − Exp[i])^2/Exp[i]$ and is calculated in the second part of Algorithm 10.5. The component of X^2 deriving from each cell is stored in an array *X2Component*; it can be helpful to inspect these values when interpreting a large total X^2. In fact this statistic is only *approximately* distributed as χ^2, and the approximation is poor if too many of the expected cell frequencies are small. A rule of thumb is that no expected frequency should be less than 5; however, this is really being rather too strict, and the topic is discussed by Snedecor and Cochran (1980). The algorithm returns a boolean variable, *SmallExpectation*, set equal to 'true' if any expected frequency falls below a chosen value, which we have set at 5 (although it is very easy to alter this choice). If expected frequencies in some classes are too small, adjacent classes should be merged to produce classes with sufficiently large expected frequencies.

A reasonable form of output is as follows, where we have inserted the algorithm variable names (in italics) in the appropriate places.

Observed	Expected	X2 component
Obs[1]	*Exp*[1]	*X2Component*[1]
Obs[2]	*Exp*[2]	*X2Component*[2]
⋮	⋮	⋮

X2 Statistic	*X2*
Degrees of freedom	(*to be inserted by user*)

The critical value needed for an $\alpha\%$ significance test may be obtained by putting $p = 1 - \alpha$ in *ChiSquaredPercentPointApprox* (Algorithm 8.13), using the appropriate degrees of freedom. The probability of the X^2-statistic, found for a particular set of data, being exceeded is obtained by input of X^2 into *ChiSquaredProb* (Algorithm 8.9); the required probability is $1 - p$, where p is the output from this algorithm.

10.5.2   Chi-squared contingency table tests

When data have been classified in the form of a two-way contingency table, the usual null hypothesis for testing is that the two classifications are independent of one another. On this assumption the expected frequencies in the cells of the table may be derived from the marginal totals of the observed table. The form of the test statistic is the same as for a goodness-of-fit test, namely $\sum (Obs[i] - Exp[i])^2 / Exp[i]$ (*ABC*, Section 18.7).

Algorithm 10.6, *FindX2ofTable* first calculates totals and expected values, and then proceeds as in the goodness-of-fit test of Section 10.5.1, a boolean variable being set if any expected frequency is small. Output could usefully contain tables of X^2 components, the observed and expected frequencies, the value of the X^2-statistic and the degrees of freedom. Critical values and *p*-values may be obtained as explained in Section 10.5.1.

```
procedure FindX2ofTable(Obs : FreqTable;
                        NoOfRows : Rows; NoOfCols : Cols;
                        var X2 : real;
                        var DF : integer;
                        var SmallExpectation : boolean;
                        var X2Component : FreqMatrix);

(* calculates table of expected values given table of      *)
(* observed values and numbers of rows and columns and then *)
(* calculates the chi-squared statistic X2. If any expected *)
(* value is less than the stated minimum the variable       *)
(* SmallExpectation is set to true.                         *)

const MinimumExpValue = 5;
```

```
var ColTotal : FreqList;
    Diff : real; (* observed minus expected frequency *)
    Exp : FreqMatrix; (* table of expected frequencies *)
    ExpectedValue: real;
    GrandTotal : integer; (* of all observed frequencies *)
    i,j : integer; (* row and column labels *)
    ObservedValue : integer;
    RowTotal : FreqList;

begin
    SmallExpectation := false;
    for j := 1 to NoOfCols do
        ColTotal[j] := 0;
    GrandTotal := 0;
    for i := 1 to NoOfRows do
    begin
        RowTotal[i] := 0;
        for j := 1 to NoOfCols do
        begin
            ObservedValue := Obs[i,j];
            RowTotal[i] := RowTotal[i] + ObservedValue;
            GrandTotal := GrandTotal + ObservedValue;
            ColTotal[j] := ColTotal[j] + ObservedValue
        end
    end;
    for i := 1 to NoOfRows do
        for j := 1 to NoOfCols do
            Exp[i,j] := RowTotal[i]*ColTotal[j]/GrandTotal;
    X2 := 0.0;
    for i := 1 to NoOfRows do
        for j := 1 to NoOfCols do
        begin
            ExpectedValue := Exp[i,j];
            if ExpectedValue < MinimumExpValue
                then SmallExpectation := true;
            Diff := Obs[i,j] - ExpectedValue;
            X2Component[i,j] := Diff*Diff/ExpectedValue;
            X2 := X2 + X2Component[i,j]
        end;
    DF := (NoOfRows-1)*(NoOfCols-1)
end; (* of FindX2ofTable *)
```

Algorithm 10.6 *FindX2ofTable*

10.6   Test data for Algorithms

Algorithm 10.3, *FindSignedRankStatistic.* (i) Input: $x = (2, 9, 7, 4, 4, 8)$; $y = (4, 8, 8, 7, 8, 6)$, $n = 6$. Output: $TMinus = 16$, $NormalDeviate = 1.15311$. (ii) Input: $x = (2, 9, 7, 4, 4, 8, 5, 5)$; $y = (4, 8, 8, 7, 8, 6, 5, 6)$; $n = 8$. Output: $TMinus = 21.5$, $NormalDeviate = 1.26773$.

Algorithm 10.4, *FindRankSumStatistic.* (i) Input: $x = (2, 3, 7, 10)$, $y = (5, 6, 9, 12, 13)$, $n1 = 4$, $n2 = 5$, Dimension of $x \geq 9$. Output: $U = 5$, $RankSum = 15$, $NormalDeviate = -1.22474$. (ii) Input: $x = (2, 3, 7, 10, 12)$, $y = (5, 6, 9, 12, 13, 15, 19)$, $n1 = 5$, $n2 = 7$, Dimension of $x \geq 12$. Output: $U = 8.5$, $RankSum = 23.5$, $NormalDeviate = -1.46160$.

Algorithm 10.5, *FindX2ofFit*. Input: $Obs = (4, 10, 8, 13, 16)$, $Exp = (5, 11, 12, 15, 8)$, $NoOfClasses = 5$. Output: $X2 = 9.89091$.

Algorithm 10.6, *FindX2ofTable*. Input: $Obs = (10, 50, 15, 25)$, $NoOfRows = 2$, $NoOfCols = 2$. Output: $X2 = 5.55556$, $DF = 1$, $X2Component = (1.66, 0.55, 2.5, 0.83)$.

10.7   Exercises

1   Write a program which:
1) generates a random sample of size k from a normal distribution with mean m and variance v;
2) calculates the sample mean and variance;
3) calculates a 95 % confidence interval for the true mean, based on the sample data (i.e. *not* using the known values m, v);
4) repeats n times, and counts the number of times t that m is actually contained in the confidence interval;
5) prints out t/n.

Run the program with $m = 10$, $v = 4$, $k = 10$, $n = 100$.
Modify step (3) to calculate a 99 % confidence interval, and run the program with the same values of m, v and k; take $n = 200$.

2   Consider a sample of pairs of values (x_i, y_i), $i = 1, 2, \ldots, n$, from a bivariate normal distribution. On the null hypothesis that the true population correlation coefficient is 0, the statistic $T = r\sqrt{(n-2)}/\sqrt{1-r^2}$ follows a t distribution with $(n-2)$ degrees of freedom. Write a program which, when given n pairs of values (x, y),
1) prints a scattergram of the data (see Algorithm 5.4);
2) calculates the sums-of-squares-and-products matrix of the data (see Algorithm 6.3);
3) prints out the values of (i) the correlation coefficient, (ii) the t-statistic described above, (iii) the number of degrees of freedom.

3   The *sign test* (*ABC*, Section 12.2.2) requires m pairs of values (x, y). It attaches a + sign to each pair in which $x > y$, a − sign when $x < y$, and ignores pairs in which $x = y$: suppose there are k of the latter type. On the null hypothesis that there is no difference between the medians of x and y in the population, the number of plus signs should be binomially distributed with the parameters $p = \frac{1}{2}$ and $n = m - k$.

Write a program which, when given m pairs of values (x, y):
1) tests whether each pair should be allocated a + or a − sign, or be ignored;
2) counts the number of plus signs, r, and the number of pairs ignored, k;
3) finds the probability p_0 of r in the appropriate binomial distribution (see Algorithm 8.1);
4) prints out m, k, r and p_0.

4   Write a program to carry out a goodness-of-fit test on a set of observations which are said to be from a binomial distribution with a given n and p. Include instructions for combining frequencies at the beginning and/or end of the distribution so as to avoid expected frequencies being too small (see Section 10.5.1).

5   Generate a set of 100 observations from a binomial distribution with $n = 12$,

$p = 0.25$ (using Algorithm 9.3) and use the program written for question 4 to make a goodness-of-fit test on them.

Print out a frequency table showing 'observed' (i.e. generated) frequencies, expected frequencies, χ^2 value and number of degrees of freedom.

6 Modify Algorithm 10.6, *FindX2ofTable*, to include a check that all frequencies input are integers and that none are negative.

7 Use Algorithm 10.5, *FindX2ofFit*, to examine
a) the hypothesis that a die is 'fair' (properly balanced) if 120 throws gave 15 ones, 25 twos, 18 threes, 15 fours, 23 fives, 24 sixes;
b) the hypothesis that a set of three coins is 'fair' if 80 tosses of the three gave no heads 10 times, one head 25 times, two heads 34 times and three heads 11 times.

8 *Fitting an exponential distribution to a set of data.* The life-times of 500 items of a particular electrical component have been summarised in the following frequency table.

Time (hours)	Frequency
0– 99	208
100–199	112
200–299	75
300–399	40
400–499	30
500–599	18
600–699	11
700–799	6

1) Write a section of program to estimate the mean of these observations. (The first part of question 10 of Exercises 6.10 will be appropriate.)
2) Continue the program by finding the expected frequencies in an exponential distribution with the same mean (i.e. whose parameter is 1/mean).
3) Carry out a goodness-of-fit test of these expected frequencies to those observed.
4) Print out the observed values in each interval, the corresponding expected values, the χ^2-value from the test and its number of degrees of freedom.

9 A geometric distribution can be computationally awkward to fit to data because its expected frequencies fall off very slowly. Construct a program for a goodness-of-fit test such that the class-intervals used in the test have approximately equal expected frequencies.

Use the program to test the hypothesis that the following observations X came from a geometric distribution with $p = \frac{1}{10}$ (see Section 8.2.3).

Value of X	1	2	3	4	5	6	7	8	9	10	11	12	13	14	15
Frequency	14	10	10	6	5	9	7	2	2	3	5	3	2	2	1

Value of X	16	17	18	19	20	21	22	23	24	25	29	39	Total
Frequency	4	3	2	1	1	3	2	1	3	2	1	1	105

10†   *Fitting a normal distribution to a set of data.* Write a program similar to that in question 8 to test the goodness-of-fit of a set of data to a normal distribution. (Note that now the variance of the data will need calculating, as well as the mean.) Use it to test whether a normal distribution can be fitted satisfactorily to the following observations on the weights of crop (grams) harvested from 200 samples of plants in a field of wheat.

Crop (g)	60–79	80–99	100–119	120–139	140–159	160–179	180–199
Frequency	6	14	67	87	19	5	2

11   (i) Write a program to test the goodness-of-fit of a set of data to a Poisson distribution, the data being given in a frequency table. Make provision for *either* (a) the mean of the distribution to be specified as input; *or* (b) the mean of the data to be calculated and used as the mean in the fitted distribution.

(ii) Write a program to test the fit of data to a Poisson distribution by the 'Index of Dispersion': if N is the total number of observations in the set of data $\{x_i\}$, then

$$\sum_{i=1}^{N} (x_i - \bar{x})^2 / \bar{x}$$ is (approximately, for large N) a χ^2 variable with $(N-1)$ degrees of freedom. (See *ABC*, Section 19.14).

12   A town records the number of motor accidents on its roads each week; at the end of each week it alters the number on boards displayed on each main road entering the town; these say 'there have been (X) accidents in this town this year'. The numbers X exhibited in successive weeks during a year were 3, 5, 6, 8, 8, 11, 12, 13, 15, 17, 20, 21, 25, 25, 26, 31, 34, 36, 38, 43, 43, 47, 49, 50, 58, 61, 62, 62, 64, 66, 66, 69, 73, 75, 80, 82, 82, 85, 86, 87, 87, 89, 91, 94, 94, 97, 102, 103, 107, 107, 109, 110.

Write a program to extract the *weekly* figures, and to test whether these follow a Poisson distribution (by either of the methods suggested in question 11).

13   Write a program which:
(a) inputs two unpaired samples of data;
(b) finds the mean and variance of each of the samples;
(c) performs an unpaired *t*-test on these data.
(See Section 10.4.1).

11 Regression

11.1 Introduction

In many investigations we can propose that there exists a linear relationship between variables, and then we use this relation to predict the value of one variable given the values of the others. For a simple two-variable situation we can do this by fitting a regression line to the measurements taken on the two variables, or some transformation of them (*ABC*, Chapter 21). When more than two variables are involved, the linear statistical model is rather more complicated (Draper and Smith, 1981). We discuss algorithms for both these situations in this chapter.

11.2 Fitting a line to pairs of values

The most common situation is when we have records of pairs of values (x, y), representing pairs of measurements on the same experimental units; we wish to fit a straight line to these when plotted on a graph, and to use the line in predicting one variable from the other. We shall assume that x is the independent variable and y the dependent variable, and shall obtain what is called the 'regression of y on x'; we will then use it to predict y from x. The estimated line takes the form

$$y = a + bx,$$

where $b = \dfrac{\Sigma(x - \bar{x})(y - \bar{y})}{\Sigma(x - \bar{x})^2}$ and $a = \bar{y} - b\bar{x}$, the sums being calculated over all the n pairs of data values available. The major part of the calculation, to derive the sums of squares and products, is carried out using Algorithm 6.3.

Corresponding to each *observed* value $y_i (i = 1, 2, \ldots, n)$ of the predicted or dependent variable there is a *fitted* value $a + bx_i$. The difference between these two values is called a **residual**:

$$r_i = y_i - a - bx_i.$$

The sizes of these residuals reflect how well the line fits the data: the smaller the residuals the better the fit. Residuals can give very useful information, and should be calculated as a matter of course when fitting a line by computer (although, unfortunately, the calculations can be prohibitive when fitting by hand). *FitLine* (Algorithm 11.1) calculates the estimates of the regression coefficients, and also the residuals. The sum of squares of residuals Σr_i^2 is calculated; $\Sigma r_i^2/(n-2)$ estimates the variance of the deviations of y about the fitted line.

A suggested form of output is as follows; expressions are shown in italics in the places where their numerical values should appear.

Fitted line is $y = a + b*x$

Coefficient	Estimate	St. Error
a	a	$StdErrora$
b	b	$StdErrorb$

Source	SS	DF	MS
Regression	$SSRegr$	1	$SSRegr$
Residual	$SSResidual$	$n-2$	$MSResidual$
Total	$SSTotal$	$n-1$	

Observed	Fitted	Residual
$y[1]$	$FittedY[1]$	$Residual[1]$
$y[2]$	$FittedY[2]$	$Residual[2]$
$\vdots$	$\vdots$	$\vdots$
$y[n]$	$FittedY[n]$	$Residual[n]$

Algorithm 11.2, *OutLine*, produces this output. The call to *OutLine* should be made immediately before the last 'end' in *FitLine*. The standard error of a is $\sigma\sqrt{\left(\dfrac{1}{n} + \dfrac{\bar{x}^2}{\Sigma(x_i - \bar{x})^2}\right)}$ and of b is $\sigma\sqrt{\left(\dfrac{1}{\Sigma(x_i - \bar{x})^2}\right)}$, where σ^2 is estimated by $\Sigma r_i^2/(n-2)$. The residual sum of squares in the analysis of variance table equals Σr_i^2 and the total sum of squares is $\Sigma(y_i - \bar{y})^2$; the regression sum of squares equals the difference between these two.

```
procedure FitLine(x,y : DataVector; n : Units;
                  var a,b,SSTotal,SSResidual : real;
                  var FittedY,Residual : DataVector);

(* finds the regression line of variate Y on variate X given *)
(* data in arrays x and y, each of size n. The equation of   *)
(* the line is of the form  Y = a + b*X. Also calculated are *)
(* the fitted Y, corresponding to each X, the sum of squares *)
(* of the residuals and the total sum of squares.            *)

var i : integer; (* loop counter *)
    SSx,SSy,SPxy : real; (* corrected sums of squares of x *)
                         (* and y and sum of products *)
    xMean,yMean : real; (* means of x and y *)


procedure FindSSandSP(x,y : DataVector; n : Units;
                      var xMean,yMean,SSx,SSy,SPxy : real);

        (*** Algorithm 6.3 ***)


begin  (* FitLine *)
   FindSSandSP(x,y,n,xMean,yMean,SSx,SSy,SPxy);
   b := SPxy/SSx;
```

```
    a := yMean - b*xMean;
    SSResidual := 0;
    for i := 1 to n do
    begin
        FittedY[i] := a + b*x[i];
        Residual[i] := y[i] - FittedY[i];
        SSResidual := SSResidual + Residual[i]*Residual[i]
    end;
    SSTotal := SSy
end; (* of FitLine *)
```

Algorithm 11.1 *FitLine*

```
procedure OutLine;

(* writes in tabular form the output from FitLine, using the *)
(* variables calculated there. The call to Outline should be *)
(* made immediately before the last 'end' in FitLine.  The   *)
(* field width and number of decimal places in the write     *)
(* statements should be adjusted to suit the data.           *)

var Hold : real; (* temporary variable *)
    MSResidual : real; (* residual mean square *)
    StdErrora,StdErrorb : real; (* standard errors of a,b *)
    SSRegr : real; (* regression sum of squares *)

begin
    MSResidual := SSResidual/(n-2);
    StdErrorb := sqrt(MSResidual/SSx);
    Hold := MSResidual*(1.0/n + xMean*xMean/SSx);
    StdErrora := sqrt(Hold);
    SSRegr := SSTotal - SSResidual;
    writeln('Fitted line is y = a + b*x');
    writeln;
    writeln('Coefficient    Estimate    St. Error');
    writeln;
    writeln('      a    ',a:9:2,'    ',StdErrora:9:2);
    writeln('      b    ',b:9:2,'    ',StdErrorb:9:2);
    writeln;
    writeln('Source       SS      DF    MS');
    writeln('Regression ',SSRegr:6:2,'     1',SSRegr:7:2);
    writeln('Residual   ',SSResidual:6:2,n-2:6,MSResidual:7:2);
    writeln('Total      ',SSTotal:6:2,n-1:6);
    writeln;writeln;
    writeln('Observed    Fitted    Residual');
    writeln;
    for i:= 1 to n do
        writeln(y[i]:9:2,FittedY[i]:9:2,Residual[i]:9:2);
end; (* of OutLine *)
```

Algorithm 11.2 *OutLine*

11.3 Residuals

The statistical model underlying the fit of a straight line to pairs of data values makes an assumption about the 'error' component of each observation, i.e. the discrepancy between the observation and its expected value: it assumes that this component is a random sample from an error distribution with mean zero and variance fixed (constant) but unknown.

In fact we cannot know these error components, but the residuals, which are the discrepancies between observed and fitted values, may be taken as a good approximation to them, particularly if we have a large amount of data. One constraint on the residuals that we should note is that their sum is zero, because of the way they are calculated. But otherwise we hope they have the properties of a random sample. If they do not, we may take it as evidence that the assumptions underlying the model are not being satisfied.

The most informative plot is usually thought to be that of residuals against fitted values. We may use *PlotScatter* (Algorithm 5.4), with *Residual* put into *y* and *FittedY* put into *x* to provide the plot. We hope to see a random scatter over the plane; any evidence of pattern suggests a deviation from assumptions. In particular, a wedge-shaped collection of points suggests that the variance is not staying constant but is changing as the level of response *y* changes. If the points lie on a curve with the residuals containing long runs of positive or negative signs, then this suggests that a curve should be fitted rather than a straight line. Odd points that are remote from the main cluster of points suggest that values may have been misrecorded.

If the data were collected at successive times, or at regular points along a strip of ground in a field experiment, the residuals should be plotted against time or distance to see whether there is evidence of association. With time data it is also worth plotting the residual of the point recorded at time t against the residual of the point recorded at time $t + 1$, for the complete set of points, to check if successive errors are independent or not. Again, *PlotScatter* (Algorithm 5.4) might be used for the plot.

Readers will find in Draper and Smith (1981) more information on the interpretation of residuals and on what to do if the assumptions fail.

11.4   Multiple regression

Sometimes we wish to predict a random variable Y not just from one variable X but from a set of variables $X_1, X_2, \ldots, X_p$. For example, we may wish to predict the mass of wood in a tree from measurements of its height, girth and age. The X variables are often called independent variables, and the Y variable the dependent variable. The procedure is called the regression of Y on $X_1, X_2, \ldots, X_p$.

We give an algorithm *FitMultiRegr* (11.3) which estimates the intercept a and the regression coefficients $b_1, b_2, \ldots, b_p$ in the regression equation

$$y = a + b_1 x_1 + b_2 x_2 + \ldots + b_p x_p,$$

using least-squares estimation. (The number of independent variables is denoted by $nVar$ in the algorithm instead of by p.) The algorithm also calculates the standard errors of the estimates, the sums of squares for an analysis of variance, and the fitted values and residuals. Multiple regression is a powerful tool but it is often misused. We recommend that the reader who has not met the technique before should read an introductory description such as is given in Snedecor and Cochran (1980), and should follow that by reading more specialised books such as Draper and Smith (1981) and Chatterjee and Price (1977).

We assume that the p independent variables $(X_1, X_2, \ldots, X_p)$ and the dependent variable Y have been measured on each of n units. The values of the independent variables are put in an array x, where each $x[j]$, $(j = 1, 2, \ldots, p)$, is an array of n elements: the values of the jth variable. The values of the dependent variable are put in a vector y with n elements. When the data are input it is probably most convenient to enter for each unit the values:

$$y \ x_1 \ x_2 \ldots x_p.$$

This might be done using the instructions:

```
for i: = 1 to n do
    for j: = 1 to p do
        read (y[i], x[j, i]);
```

Note that we write $x[j, i]$ and not $x[i, j]$. We require the ith element of the array $x[j]$ which we can write $x[j][i]$ or, more conveniently, as $x[j, i]$.

The algorithm begins by making y the $(p + 1)$th column of the matrix x, and then enters the procedure *FindMultiSP* (Algorithm A.3), which is in Appendix A. This procedure calculates the vector of means and the matrix of sums of squares and products. (Note that since the matrix is symmetric, only the upper triangle need be stored. Moreover, for more efficient computation, the matrix is stored in a one-dimensional array *SP* which is the upper triangle, read by rows, of the matrix; it is convenient to declare this as type *MatrixArray* (see page xi). Other types introduced (*Variates, DataMatrix*) are also defined on page xi.) The procedure is a generalisation to many variables of Algorithm 6.3, *FindSSandSP*.

```
procedure FitMultiRegr(y : DataVector; x : DataMatrix;
                n : Units; nVar : Variates;
                var a,SSResidual,SSTotal : real;
                var b,xMean : Vector;
                var FittedY,Residual : DataVector;
                var SPInverse : MatrixArray;
                var Nullity : integer);

(* calculates the regression of y on the variables x, given  *)
(* n sets of observations. The regression coefficients       *)
(* appear in b and the intercept in a. Also calculated are   *)
(* the fitted values and the residuals, the residual sum of  *)
(* squares, the total sum of squares and the inverse of the  *)
(* sum of squares and products matrix. Nullity gives a count *)
(* of the number of zero diagonal elements in the inverse.   *)
(* The data matrix should be declared not smaller than n by  *)
(* (nVar + 1), where nVar is the number of x variables.      *)

var i,j : integer; (* counters *)
    nVarPlusOne : integer; (* nVar + 1 *)
    RHS : Vector ; (* right hand side vector of normal eqns  *)
    RowMark : integer; (* label *)
    SP : MatrixArray; (* sums of squares matrix,in vector *)
    Sum1,Sum2 : real; (* temporary accumulators *)
    yMean : real; (* mean of the dependent variable *)
```

```
procedure FindMultiSP(x : DataMatrix; n : Units; p : Variates;
                      var Mean : Vector;
                      var SP : MatrixArray);

          (***   Algorithm A.3   ***)


procedure LinearSolver(A : MatrixArray; h : Vector;
                       NoOfEqns : integer;
                       var x : Vector;
                       var AInverse : MatrixArray;
                       var Nullity : integer);

          (***   Algorithm A.4   ***)




begin   (*   FitMultiRegr   *)

    nVarPlusOne := nVar + 1;
    x[nVarPlusOne]:= y;

    FindMultiSP(x,n,nVarPlusOne,xMean,SP);

    RowMark := nVar*nVarPlusOne div 2;
    for i :=1 to nVar do
        RHS[i]:= SP[RowMark+i];
    SSTotal := SP[RowMark+nVarPlusOne];
    yMean := xMean[nVarPlusOne];

    LinearSolver(SP,RHS,nVar,b,SPInverse,Nullity);

    Sum1 := 0;
    for i := 1 to nVar do
        Sum1 := Sum1 + b[i]*xMean[i];
    a := yMean - Sum1;
    Sum2 := 0;
    for i := 1 to n do
    begin
        Sum1 := a;
        for j := 1 to nVar do
            Sum1 := Sum1 + b[j]*x[j,i];
        FittedY[i] := Sum1;
        Residual[i] := y[i] - Sum1;
        Sum2 := Sum2 + Residual[i]*Residual[i]
    end;
    SSResidual := Sum2

end; (*   of FitMultiRegr   *)
```

Algorithm 11.3 *FitMultiRegr*

The array *SP*, when written in matrix form, may be considered as partitioned into submatrices:

$$\begin{bmatrix} \mathbf{A} & \mathbf{h} \\ \mathbf{h}^{\mathrm{T}} & d \end{bmatrix}.$$

If we write $m_j (j = 1, 2, \ldots, p)$ for the mean of x_j and m_{p+1} for the mean of y, then **A** is a $p \times p$ matrix whose j, kth element is $\sum_{i=1}^{n} (x_{ji} - m_j)(x_{ki} - m_k)$, **h** is a vector whose jth element is $\sum_{i=1}^{n} (x_{ji} - m_j)(y_i - m_{p+1})$ and d is $\sum_{i=1}^{n} (y_i - m_{p+1})^2$. The normal equations, whose solution is the vector of regression coefficients **b**, are given by

$$\mathbf{Ab} = \mathbf{h}.$$

Algorithm 11.3, *FitMultiRegr*, uses another procedure *LinearSolver* (Algorithm A.4) to solve the equations using a Cholesky decomposition; the solutions are put in the array *b*. The procedure also calculates the inverse matrix $\mathbf{A}^{-1}$: this is needed for calculating the standard errors of the estimates. The matrices are all stored as one-dimensional arrays.

The algorithm proceeds to calculate the intercept *a*, the fitted values, called *FittedY* and the residuals. Earlier in the algorithm, the sum of squares of deviations of *y*, which is called the **total sum of squares** in the analysis of variance, is calculated. The **residual sum of squares** is calculated by summing the squares of the residuals.

```
procedure OutMultiRegr(y : DataVector; x : DataMatrix;
                       n : Units; nVar : Variates;
                       var a,SSResidual,SSTotal : real;
                       var b,xMean : Vector;
                       var FittedY,Residual : DataVector;
                       var SPInverse : MatrixArray);

(* writes, in tabular form, the output from FitMultiRegr   *)
(* using the variables calculated there. The field width    *)
(* and number of decimal places in the write statements(e.g. *)
(* :9:2) should be adjusted to suit the data.               *)

var Sum : real; (* temporary variable *)
    k : integer; (* label *)
    MSResidual : real; (* residual mean square *)
    StdErrora : real; (* standard error of a *)
    StdErrorb : real; (* standard error of a component of b *)
    SSRegr : real; (* regression sum of squares *)

begin
    MSResidual := SSResidual/(n-nVar-1);
    SSRegr := SSTotal - SSResidual;
    Sum := SPInverse[1]*xMean[1]*xMean[1];
    for i := 2 to nVar do
    begin
        for j := 1 to i-1 do
        begin
            k := (i*(i-1)) div 2 + j;
            Sum := Sum + 2*SPInverse[k]*xMean[i]*xMean[j]
        end;
        Sum := Sum + SPInverse[k+1]*xMean[i]*xMean[i]
    end;
    StdErrora := sqrt(MSResidual*(Sum+1.0/n));
    writeln('Coefficient   Estimate   St. Error');
    writeln;
    writeln('      a   ',a:9:2,'      ',StdErrora:9:2);
```

```
for i := 1 to nVar do
begin
    k := (i*(i+1)) div 2;
    StdErrorb := sqrt(MSResidual*SPInverse[k]);
    writeln('      b(',i:1,')',b[i]:9:2,'        ',
                                        StdErrorb:9:2)
end;
writeln;
writeln('Source          SS      DF   MS');
writeln('Regression    ',SSRegr:6:2,nVar:6,
                                SSRegr/nVar:7:2);
writeln('Residual      ',SSResidual:6:2,n-nVar-1:6,
                                MSResidual:7:2);
writeln('Total         ',SSTotal:6:2,n-1:6);
writeln;writeln;
writeln('Observed    Fitted     Residual');
writeln;
for i:= 1 to n do
    writeln(y[i]:9:2,FittedY[i]:9:2,Residual[i]:9:2);
end; (*  of OutMultiRegr  *)
```

Algorithm 11.4 *OutMultiRegr*

Algorithm 11.4, *OutMultiRegr*, provides an appropriate output. The form of the list of estimates and of the analysis of variance is shown in the output layout given below; a list of residuals is also printed. The variances and covariances of the regression coefficients are given by $\sigma^2 \mathbf{A}^{-1}$, where σ^2 is estimated by the Residual Mean Square. Thus the standard error of b_i is estimated by σ times the ith diagonal element of $\mathbf{A}^{-1}$. The standard error of a is rarely of interest and is not included in the layout. If it should be required it may be calculated from the expression $\sigma\sqrt{(1/n + \mathbf{m}^T \mathbf{A}^{-1} \mathbf{m})}$, where $\mathbf{m}$ is the vector of variate means, $\mathbf{A}^{-1}$ is the inverse matrix described above and σ is replaced by sqrt(*MSResidual*).

Coefficient	Estimate	St. Error
a	*a*	*StdErrora*
b(1)	*b*[1]	*StdErrorb*[1]
b(2)	*b*[2]	*StdErrorb*[2]
⋮	⋮	⋮

Source	SS	DF	MS
Regression	*SSRegr*	*nVar*	*SSRegr/nVar*
Residual	*SSResidual*	*n − nVar − 1*	*MSResidual*
Total	*SSTotal*	*n − 1*	

11.5   Test data for Algorithms

Algorithm 11.1, *FitLine*. Input: $n = 5$, $x = (3, 6, 7, 11, 13)$, $y = (4, 7, 5, 8, 11)$, $xMean = 8$, $yMean = 7$, $SSx = 64$, $SPxy = 40$, $SSy = 30$. Output: $a = 2$, $b = 0.625$, *FittedY*

$= (3.875, 5.750, 6.375, 8.875, 10.125)$, *Residual* $= (0.125, 1.250, -1.375, -0.875, 0.875)$, *SSTotal* $= 30$, *SSResidual* $= 5$.

Algorithm 11.3, *FitMultiRegr*. Input: $n = 5, p = 2, y = (4, 7, 5, 8, 11)$, $x = (3, 10), (6, 8)$, $(7, 5), (11, 4), (13, 3)$. Output: $a = -8.7\dot{3}$, $b = (1.2\dot{6}, 0.9\dot{3})$. *SSResidual* $= 1.7\dot{3}$, *SSTotal* $= 30$. *FittedY* $= (4.40, 6.3\dot{3}, 4.80, 8.9\dot{3}, 10.5\dot{3})$. *Residual* $= (-0.40, 0.66, 0.20, -0.9\dot{3}, 0.4\dot{6})$.

11.6 Exercises

1 The following data give the reaction times of 10 men, of various ages, to a visual stimulus in a psychological experiment; X is age (in years) and Y is reaction time (in milliseconds):

X	37	35	41	43	42	50	49	54	60	65
Y	190	197	205	210	218	226	228	230	234	240

Apply Algorithms 11.1 and 11.2 to calculate the regression line of Y on X and print out the results, including residuals.

2 Write a program which
i) inputs n data in pairs (x_i, y_i);
ii) plots a scattergram (Algorithm 5.4);
iii) calculates the SS and SP values (Algorithm 6.3);
iv) prints the correlation coefficient (see Section 6.6);
v) calculates the regression line of Y on X (Algorithm 11.1);
vi) prints the results of (v) (Algorithm 11.2).

Check the program using the test data for Algorithm 11.1 (Section 11.5).

3 Write a program to:
i) generate n pairs of data (x, y) according to the model

$$y_i = a + bx_i + e_i \qquad (i = 1, 2, \ldots, n)$$

where $x_i = i$ and e_i is a random sample from a normal distribution with mean 0 and variance σ^2;
ii) estimate the values of the intercept a and slope b in the regression line of Y on X;
iii) repeat this k times and study the distributions of the estimates of a and b.

Run the program with $n = 20, a = 5, b = 3, \sigma^2 = 1$ and $k = 100$. Hence verify (within the limits of this type of investigation) that the least squares estimators of a and b are unbiased, and that their standard errors are as quoted in Section 11.2.

4 Extend the program from question 1 to plot out the observed values of Y at each X, and also (on the same diagram) the values of Y predicted from the fitted regression line.

5† Use Algorithms 11.3, *FitMultiRegr* and 11.4, *OutMultiRegr*, to examine the relation between W, to be taken as dependent variable, and H, A, taken as independent variables, in the data set of Appendix B. Carry out the regression separately for males and females, and base each regression on a sample of 20 chosen at random from the whole data set.

6†   Continuing question 5, carry out regression analyses, using Algorithms 11.1 and 11.2, of (i) W on H, and (ii) W on A, for males and for females separately. Hence examine whether predicting W from both independent variables together, in a multiple regression, gives better results than prediction using only one of the independent variables. (Measure this by comparing the sums of squares for regression with one independent variable and with two independent variables: the difference, with one degree of freedom, is the improvement due to including the second variable.)

12 Analysis of variance

12.1 Introduction

Data collected by means of experiments are usually analysed within the framework of an Analysis of Variance. This technique splits the total variability among all the observations into components. One of these components will correspond to the differences between the effects of the experimental treatments on the experimental units, which may be, for example, plants, animals, or items of industrial output from factories. Often the units are grouped into blocks, each of which contains relatively uniform units; a component in the analysis will then correspond to differences between these blocks. The final component is the residual variability among units. When carrying out the analysis, the treatment variation is compared with the residual variability, and if it is about the same size we may conclude that the treatments are all having a similar effect; if, however, treatment variation is significantly larger than residual variation we conclude that there are real differences between the effects of the treatments.

The actual quantity which we split into components is the **total sum of squares**, $\Sigma(x_i - \bar{x})^2$, of the differences between each individual observation x_i and the mean of the whole data set $\bar{x}$. The components are called the **treatment sum of squares**, the **block sum of squares** (if blocks have in fact been used in the experiment) and the **residual sum of squares**. There are certain 'overall' tests which can be made in an analysis of variance, but the precise information on the performance of each experimental treatment is best shown in a table of means. A table of residuals is also useful, to check (as in regression analysis) whether the assumptions underlying the analysis are being satisfied.

In this chapter we present algorithms for analysing data from the two most common types of designed experiment: completely randomised and randomised block designs. Similar forms of analysis of variance are sometimes used with non-experimental data, so we shall name the analysis of variance algorithms according to the structure of the data: there may be **one-way** (single classification) data, as in a completely randomised design, or **two-way** (cross-classification) data as in a randomised block design. A description of the collection and analysis of experimental data, and a discussion of the principles of the analysis of variance, will be found in *ABC* (Chapters 11 and 22) and in Clarke (1980).

12.2 Input of data

When a set of data arises from a simple designed experiment, each observation can be labelled by its treatment and replicate numbers. In a randomised block design, replicate *j* of each treatment comes from block *j*, this collection (block) of experimental units, or plots, having been chosen to be as similar as possible. With a completely randomised design, however, replicate *j* of treatment *i* is not related at all to replicate *j* of treatment *k*. For each plot there are three items of information: (1) the treatment number, or name, (2) the replicate number, (3) the value of the variate, e.g. yield. These items of information are not all of the same variable type so a **record** is an appropriate Pascal data structure. The complete set of data may be regarded as an array of records, with one record per plot. The two procedures for analysis of variance (Algorithms 12.2 and 12.4) assume the following type declarations:

```
const MaxSampleSize = 100;   (* These constants should *)
      MaxRep = 20;           (* be adjusted to suit    *)
      MaxTr = 5;             (* the data.              *)

type Reps = 1..MaxRep;
     Trs = 1..MaxTr;
     Units = 1..MaxSampleSize;
     LineOfData = record
                     TrNo : Trs;
                     RepNo : Reps;
                     Value : real
                  end;
     DataRecords = array[Units] of LineOfData;
     DataVector = array[Units] of real;
     List = array[Trs] of integer;
     Vector = array[Trs] of real;
```

A variable *DataSet* of type *DataRecords* is declared.

So as to reduce copying errors, it is desirable to input data into a computer directly from experimental record sheets, without any intermediate stage. We therefore recommend that the input algorithm should not require the observations to be entered

```
procedure InputAnova(var DataSet : DataRecords;var n :  Units;
                     var NoOfTrs : Trs);

(* Input of data, for Anova one-way or two-way, in the form *)
(* of records. Treatments and replicates must be numbered   *)
(* from one upwards.                                         *)

var i : integer; (* counter *)

begin
    write('Number of units       ');readln(n);
    write('Number of treatments  ');readln(NoOfTrs);
    for i :=1 to n do
    with DataSet[i] do
    begin
        writeln;
        write('Treatment   ');readln(TrNo);
        write('Replicate   ');readln(RepNo);
        write('Value       ');readln(Value)
    end
end; (* of InputAnova *)
```

Algorithm 12.1 *InputAnova*

in any particular order (such as by blocks or by treatments), but instead allow each observation to be identified by its treatment and replicate numbers. Algorithm 12.1, *InputAnova*, is a simple procedure that is adequate for the input of small sets of data to be analysed by one-way or two-way analysis of variance.

Often a number of variates will be recorded at the same time on each experimental unit, e.g. total weight of apples from a tree, weight of grade one apples and total number of apples. It is easy to extend Algorithm 12.1 to input all these data simultaneously, each variate being put into its own separate data vector. We recommend that all but small sets of data be input by an algorithm which incorporates adequate checks as the data are entered (see Chapter 13).

12.3 One-way analysis of variance

The first part of Algorithm 12.2, *AnovaOneWay*, calculates the treatment totals and means, and in the second part these are used to derive the sums of squares. The treatments sum of squares is calculated from the differences between the treatment means and the grand mean, and the total sum of squares from the differences between individual observations and the grand mean. The residual sum of squares equals the total sum of squares minus the treatments sum of squares. Each **plot residual**, which is the difference between the observation (or 'yield') on that plot and the mean of all plots receiving the same treatment, is also calculated. We give a procedure *OutputAnovaOne* (Algorithm 12.3) which should follow *AnovaOneWay*.

```
procedure AnovaOneWay(DataSet:DataRecords; n:Units ;
                      NoOfTrs:Trs;
                      var SSResidual,SSTreatments : real;
                      var Residual : DataVector;
                      var TrMean : Vector;
                      var RepsOfTr : List;
                      var GrandMean : real);

(* calculates one-way analysis of variance from records of    *)
(* each of n units. The number of treatments must be given.   *)
(* The procedure calculates the number of replicates of each  *)
(* treatment,  the treatment means, the grand mean,the plot   *)
(* residuals, the sum of squares of treatments and of         *)
(* residuals.                                                  *)

var Dev : real; (* of a quantity from a mean *)
    i,j : integer; (* loop counters and labels *)
    SSTotal : real; (* total sum of squares *)
    Total : real; (* current total of observations *)
    TrSum : Vector; (* current treatment sum *)


begin  (* AnovaOneWay *)

    (* summation of values *)
    Total := 0.0;
    for i := 1 to NoOfTrs do
    begin
        TrSum[i] := 0.0;
        RepsOfTr[i] := 0
```

```
    end;
    for i := 1 to n do
        with DataSet[i] do
        begin
            TrSum[TrNo] := TrSum[TrNo] + Value;
            RepsOfTr[TrNo] := RepsOfTr[TrNo] + 1
        end;
    for i := 1 to NoOfTrs do
    begin
        TrMean[i] := TrSum[i]/RepsOfTr[i];
        Total := Total + TrSum[i]
    end;
    GrandMean := Total/n;

    (* calculation of sums of squares *)
    SSTreatments := 0.0;
    SSTotal := 0.0;
    for i:= 1 to NoOfTrs do
    begin
        Dev := TrMean[i] - GrandMean;
        SSTreatments := SSTreatments + RepsOfTr[i]*Dev*Dev;
    end;
    for i := 1 to n do
    with DataSet[i] do
    begin
        Dev := Value - GrandMean;
        SSTotal := SSTotal + Dev*Dev;
        Residual[i] := Value - TrMean[TrNo]
    end;
    SSResidual := SSTotal - SSTreatments
end; (* of AnovaOneWay *)
```

Algorithm 12.2 *AnovaOneWay*

Output AnovaOne presents the results of the analysis in three tables: (1) analysis of variance, (2) table of treatment means, (3) table of residuals. The first two tables take the following form (in which we have inserted in italics the names of the expressions in the positions that the numerical values would occupy in an actual output).

Source	SS	DF	MS
Treatments	*SSTreatments*	*DFTreatments*	*SSTreatments/DFTreatments*
Residual	*SSResidual*	*DFResidual*	*MSResidual*
Total	*SSTotal*	$n-1$	

F(*DFTreatments, DFResidual*) = *F*

Treatment	Mean	St. error
1	*TrMean*[1]	*sqrt (MSResidual/RepsofTr*[1])
2	*TrMean*[2]	*sqrt (MSResidual/RepsofTr*[2])
⋮	⋮	⋮

Overall mean *GrandMean*

```
procedure OutputAnovaOne(DataSet:DataRecords; n:Units ;
                         NoOfTrs:Trs;
                         SSResidual,SSTreatments : real;
                         Residual : DataVector;
                         TrMean : Vector;
                         RepsOfTr : List);

(* produces from the quantities calculated in AnovaOneWay    *)
(* (1) Anova table, (2) table of means, (3) table of         *)
(* residuals.  The field width and number of decimal places  *)
(* (e.g :6:2) in the write statements should be adjusted to   *)
(* suit the data.                                            *)

var DFResidual,DFTreatments : integer; (* residual,treatment *)
                                       (* degrees of freedom *)
    F : real; (* value of the F statistic *)
    i,j : integer; (* loop counters *)
    MSResidual : real; (* residual mean square *)
    SSTotal : real; (* total sum of squares *)

begin
    DFTreatments := NoOfTrs - 1;
    DFResidual := n - NoOfTrs;
    MSResidual := SSResidual/DFResidual;
    SSTotal := SSTreatments + SSResidual;
    F := (SSTreatments/DFTreatments)/MSResidual;
    writeln('Source           SS        DF      MS');
    writeln('Treatments    ',SSTreatments:6:2,DFTreatments:6,
                            SSTreatments/DFTreatments:7:2);
    writeln('Residual      ',SSResidual:6:2,DFResidual:6,
                            MSResidual:7:2);
    writeln('Total         ',SSTotal:6:2,n-1:6);
    writeln;writeln;
    writeln('Treatment     Mean     St.Error');writeln;
    for i := 1 to NoOfTrs do
        writeln(i:6,TrMean[i]:11:2,
                    sqrt(MSResidual/RepsOfTr[i]):10:2);
    writeln;
    writeln('F(',DFTreatments:2,',',DFResidual:2,
                            ') = ',F:6:2);
    writeln;writeln;
    writeln('Treatment   Replicate   Observation   Residual');
    writeln;
    for i := 1 to n do
    with DataSet[i] do
        writeln(TrNo:6,RepNo:11,Value:15:2,Residual[i]:14:2);
end; (* of OutputAnovaOne *)
```

Algorithm 12.3 *OutputAnovaOne*

The third table lists observations and corresponding residuals, the observations being in the order they were entered. It is usually convenient to place together residuals corresponding to the same treatment so that, for example, it is easy to check if variability is the same in all treatments. We therefore recommend using *OrderRecords* (Algorithm A.5, page 160) immediately after *InputAnova* (Algorithm 12.1). This algorithm re-orders the records (and hence the residuals) so that all Treatment 1 records come first, in increasing order of replicate number; then Treatment 2 records, in increasing order of replicate number; etc. Other methods of ordering may be desirable in some applications.

The plot residuals indicate how well the statistical model underlying the analysis of variance fits the data: the smaller the residuals the better the fit. They may be analysed in a similar way to the residuals produced in regression analysis (Section 11.3), in order to check how well the underlying assumptions appear to be satisfied.

12.4   Two-way analysis of variance

Algorithms 12.4, *AnovaTwoWay*, and 12.5, *OutputAnovaTwo*, are very similar to the one-way Anova algorithms. Treatment and block (= replicate) totals, and the grand mean, are obtained in the first part of *AnovaTwoWay*; these are used to calculate the Treatment, Block and Residual sums of squares in the second part. *OutputAnovaTwo* then produces (1) the analysis of variance table, (2) a table of treatment means, (3) a table of residuals.

```
procedure AnovaTwoWay(DataSet : DataRecords; n : Units;
                      NoOfTrs : Trs;
                      var SSBlocks,SSResidual : real;
                      var SSTreatments : real;
                      var Residual : DataVector;
                      var BlockMean,TrMean : Vector;
                      var GrandMean : real );

(* calculates the two-way analysis of variance from      *)
(* observations entered in an array of records(DataSet).  *)
(* The procedure calculates the Sum of Squares of Blocks, *)
(* Treatments and Residuals.  It also finds the residuals, *)
(* the treatment means, the block means and the grand mean. *)

var BlockSum : Vector;
    BlockDev : real; (* block mean minus grand mean *)
    FittedValue : real;
    i,j,k : integer; (* loop counters and labels *)
    SS : real; (* current value of a sum of squares *)
    NoOfBlocks : integer; (* number of blocks *)
    Total : real; (* current total of observations *)
    TrDev : real; (* treatment mean minus grand mean *)
    TrSum : Vector; (* sum of observations of a treatment *)
    Value : real; (* temporary store for an observation *)

begin (* AnovaTwoWay *)

    (* summation of values *)
    NoOfBlocks := n div NoOfTrs;
    Total := 0;
    for j := 1 to NoOfTrs do
        TrSum[j] := 0;
    for k := 1 to NoOfBlocks do
        BlockSum[k] := 0;
```

```
    for i := 1 to n do
    with DataSet[i] do
    begin
        TrSum[TrNo] := TrSum[TrNo] + Value;
        BlockSum[RepNo] := BlockSum[RepNo] + Value;
        Total := Total + Value
    end;
    GrandMean := Total/n;

    (* calculation of sums of squares *)
    SS := 0;
    for j := 1 to NoOfTrs do
    begin
        TrMean[j] := TrSum[j]/NoOfBlocks;
        TrDev := TrMean[j] - GrandMean;
        SS := SS + TrDev*TrDev
    end;
    SSTreatments := NoOfBlocks*SS;
    SS := 0;
    for k := 1 to NoOfBlocks do
    begin
        BlockMean[k] := BlockSum[k]/NoOfTrs;
        BlockDev := BlockMean[k] - GrandMean;
        SS := SS + BlockDev*BlockDev
    end;
    SSBlocks := NoOfTrs*SS;
    SS :=0;
    for i := 1 to n do
    with DataSet[i] do
    begin
        FittedValue := TrMean[TrNo]+BlockMean[RepNo]-GrandMean;
        Residual[i] := Value - FittedValue;
        SS := Residual[i]*Residual[i];
    end;
    SSResidual := SS;

end; (* of AnovaTwoWay *)
```

Algorithm 12.4 *AnovaTwoWay*

```
procedure OutputAnovaTwo(DataSet : DataRecords; n : Units;
                         NoOfTrs : Trs;
                         SSBlocks,SSResidual : real;
                         SSTreatments : real;
                         Residual : DataVector;
                         TrMean : Vector;
                         GrandMean : real );


(* from quantities calculated in AnovaTwoWay, produces  *)
(* (1) Anova table, (2) table of means, (3) table of    *)
(* residuals. The field width and the number of decimal  *)
(* places(e.g.:6:2) in the write statements should be    *)
(* adjusted to suit the data.                            *)

var DFBlocks,DFResidual,DFTreatments : integer;
        (* block,residual and treatment degrees of freedom *)
```

```
    F1,F2 : real; (* values of the F statistic *)
    i,j : integer; (* counter,label *)
    MSResidual : real; (* residual mean square *)
    NoOfBlocks : integer; (* number of blocks *)
    SSTotal : real; (* total sum of squares *)

begin
    NoOfBlocks := n div NoOfTrs;
    DFBlocks := NoOfBlocks - 1;
    DFTreatments := NoOfTrs - 1;
    DFResidual := DFBlocks*DFTreatments;
    MSResidual := SSResidual/DFResidual;
    SSTotal := SSTreatments + SSBlocks + SSResidual;
    F1 := (SSTreatments/DFTreatments)/MSResidual;
    F2 := (SSBlocks/DFBlocks)/MSResidual;
    writeln('Source            SS        DF      MS');
    writeln('Treatments    ',SSTreatments:6:2,DFTreatments:6,
                             SSTreatments/DFTreatments:7:2);
    writeln('Blocks        ',SSBlocks:6:2,DFBlocks:6,
                             SSBlocks/DFBlocks:7:2);
    writeln('Residual      ',SSResidual:6:2,DFResidual:6,
                             MSResidual:7:2);
    writeln('Total         ',SSTotal:6:2,n-1:6);
    writeln;
    writeln('Treatments  F(',DFTreatments:2,',',DFResidual:2,
                             ') = ', F1:6:2);
    writeln('Blocks      F(',DFBlocks:2,',',DFResidual:2,
                             ') = ', F2:6:2);
    writeln;writeln;
    writeln('Treatment     Mean');writeln;
    for i := 1 to NoOfTrs do
        writeln(i:6,TrMean[i]:11:2);
    writeln;
    writeln('Overall Mean            ',GrandMean:6:2);
    writeln('S.E of Treatment Mean  ',
                sqrt(MSResidual/NoOfBlocks):6:2);
    writeln;writeln;
    writeln('Treatment   Block   Observation   Residual');
    writeln;
    for i := 1 to n do
    with DataSet[i] do
        writeln(TrNo:6,RepNo:9,Value:15:2,Residual[i]:14:2);
end; (* of OutputAnovaTwo *)
```

Algorithm 12.5 *OutputAnovaTwo*

The output of the analysis of variance and the table of treatment means takes the following form, where the expressions are shown in italics in the positions that their numerical values would occupy in the actual output.

Source	SS	DF	MS
Treatments	*SSTreatments*	*DFTreatments*	*SSTreatments/DFTreatments*
Blocks	*SSBlocks*	*DFBlocks*	*SSBlocks/DFBlocks*
Residual	*SSResidual*	*DFResidual*	*MSResidual*
Total	*SSTotal*	$n-1$	

Treatments   F(*DFTreatments/DFResidual*) = *F*1
Blocks   F(*DFBlocks/DFResidual*) = *F*2

Treatment   Mean

   1   *TrMean*[1]
   2   *TrMean*[2]
   ⋮   ⋮

Overall mean   *GrandMean*

S.E of treatment mean *sqrt*(*MSResidual/NoOfBlocks*)

Again we recommend the use of *OrderRecords* (Algorithm A.5) immediately after *InputAnova*, so that the table of residuals lists the values in treatment and block order. Ordering of records for two-way analysis of variance alone can be done much more simply (see Exercises 12.6, question 10).

12.5 Test data for Algorithms

Algorithms 12.1, *InputAnova*, and 12.2, *AnovaOneWay*. Input: $n = 14$, *NoOfTrs* = 3. Values (to be preceded by treatment and replicate numbers): Tr1: 1, 4, 4; Tr2: 3, 6, 7, 9, 10; Tr3: 6, 8, 10, 11, 12, 13. Output: *RepsOfTr* = (3, 5, 6), *SSResidual* = 70, *SSTreatments* = 99.43, *GrandMean* = 7.43, *TrMean* = (3, 7, 10), *Residual* = $(-2, 1, 1, -4, -1, 0, 2, 3, -4, -2, 0, 1, 2, 3)$.
Algorithms 12.1, *InputAnova*, and 12.4, *AnovaTwoWay*. Input: $n = 12$, *NoOfTrs* = 3. Values (in replicate order within treatments; precede by treatment and replicate number): Tr1: 1, 2, 3, 6; Tr2: 2, 6, 7, 9; Tr3: 8, 10, 11, 15. Output: *SSResidual* = 4, *SSTreatments* = 130.67, *SSBlocks* = 62, *GrandMean* = 6.66, *TrMean* = (3, 6, 11), *BlockMean* = (3.66, 6, 7, 10), *Residual* = $(1, -\frac{1}{3}, -\frac{1}{3}, -\frac{1}{3}, -1, \frac{2}{3}, \frac{2}{3}, -\frac{1}{3}, 0, -\frac{1}{3}, -\frac{1}{3}, \frac{2}{3})$.

12.6 Exercises

1 Write an input algorithm for experimental data, on the lines suggested in Section 12.2, capable of handling several variates at one time.

2 Construct a program which analyses data from a completely randomised design, and which
  i) inputs experimental data (Algorithm 12.1);
  ii) carries out a one-way analysis of variance on the data (Algorithm 12.2);
  iii) prints out the analysis of variance table, a table of means and a table of residuals (Algorithm 12.3).

Use this complete program with the following data.

Treatment	Replicate	Value	Treatment	Replicate	Value
1	1	24	2	3	37
2	1	46	3	4	41
1	2	18	2	4	50
1	3	18	2	5	44
3	1	32	3	5	36
3	2	30	3	6	28
1	4	29	1	7	15
2	2	39	2	6	45
1	5	22	2	7	30
1	6	17	3	7	27
3	3	26			

3   Construct a program, similar to that described in question 2, but for a two-way analysis of variance using Algorithms 12.1, 12.4 and 12.5.
   Use this program with the following data.

Treatment	Block	Value	Treatment	Block	Value
1	1	68	4	3	103
2	1	71	1	3	77
3	1	54	5	3	88
2	2	78	2	3	74
3	2	67	3	3	65
4	1	95	1	4	59
5	1	73	2	4	70
1	2	82	4	4	90
4	2	116	3	4	54
5	2	85	5	4	76

4   (a) The residual in an observation from a completely randomised experiment is the difference between the observation and the mean of all observations on the same treatment. Use Algorithm 5.4, *PlotScatter*, to plot the residuals against the treatment means (see Section 11.3). Use the data in question 2 above.
(b) The residual in an observation from a randomised block experiment is $r_{ij} = x_{ij} - m - t_i - b_j$; m is the mean of all observations in the experiment, $m + t_i$ is the mean for treatment i, and $m + b_j$ is the mean for block j. The corresponding fitted value is $m + t_i + b_j$. Plot the residuals against the fitted values. Use the data in question 3 above.

5   Write a program to:
 i) generate n observations according to the model

$$x_{ij} = m + t_i + e_{ij},$$

where $t_i = 4i$ and e_{ij} is a random sample from a normal distribution with mean 0 and error variance σ^2; the values taken by j are 1 to n_i, where the $\{n_i\}$ need not all be the same (but must add to n);

ii) carry out a one-way analysis of variance on the observations generated;

iii) compare the treatment means and the error variance estimated from the analysis with those used in the generation;

iv) repeat this k times and study the distribution of the estimates of the error variance.

Run the program with $i = 1$ to 4, $n_1 = 7$, $n_2 = 4$, $n_3 = 5$, $n_4 = 6$, $\sigma^2 = 3.24$, $m = 1$, $k = 50$.

6 Write a program to

i) generate n observations according to the model

$$x_{ij} = m + t_i + b_j + e_{ij},$$

where $t_i = 3i$ and $b_j = j + 5$, and e_{ij} is a random sample from a normal distribution with mean 0 and error variance σ^2; the values of i run from 1 to t_0 and of j from 1 to r_0, where $r_0 t_0 = n$;

ii) carry out a two-way analysis of variance on the observations generated;

iii) compare the treatment means and the error variance estimated from the analysis with those used in the generation;

iv) repeat this k times and study the distribution of the estimates of the error variance.

Run the program with $i = 6$, $j = 4$, $m = 5$, $\sigma^2 = 2.25$, $k = 50$.

7 It is sometimes useful to analyse a transformed function of the observations. Amend the program for one-way analysis of variance so that it can also operate on the logarithms of the original observations, and print out an appropriate message when it has done so.

Use the amended program to analyse the following experimental data, carrying out a log transformation first.

Treatment A: 12.1, 16.3, 11.8, 14.5, 13.7, 15.2
Treatment B: 17.5, 25.4, 22.6, 31.3, 42.0, 28.0, 37.5, 34.8
Treatment C: 50.3, 39.1, 47.4, 62.6, 48.3, 57.7, 35.9

8† Use Algorithm 12.2, *AnovaOneWay*, to compare the mean of H (Student Data Set, Appendix B) for males with its mean for females. Repeat for W, R, and C of Appendix B.

9 In a two-way analysis of variance, the number of replicates is constant. This fact allows the procedure *OrderRecords* to be made much simpler. If r is the number of replicates of each treatment then the 'place' of the jth replicate of the ith treatment in *DataSet* is

$$\text{place} = (i-1) * r + j.$$

Modify Algorithm A.5, *OrderRecords*, to do this.

13 Data input and use of files

13.1 Introduction

The prime task of Statistics is the analysis of data, and any analysis of data by computer begins with putting the data into the computer. It is vital that programs operate on correct data so the entry of data should be made convenient, to reduce errors, and data entered must always be checked so that errors may be detected and removed before data analysis begins. To some extent, convenience of entry and correction is a matter of personal opinion. The reader would do best to design particular input procedures for his particular input tasks. In the first part of this chapter we offer some ideas to help the reader do this.

A great deal of effort can go into entering and checking data; we recommend that all but very small sets of data should be stored in a file. We discuss the use of files for statistical work in the second part of the chapter.

13.2 Counting data on input

It may be more convenient to allow the computer to count how many data are entered rather than expect the user to make the count before he enters the data. This can be done by typing a **marker** or **sentinel** value at the end of the data. Algorithm 13.1, *Input AndCount Data*, has been derived from the simple input procedure (Algorithm 4.1) to allow this form of entry.

A marker of '−1', as used in the algorithm, is convenient if all the data are positive values but could be misleading if there were negative values; an alternative numerical marker is a very large number such as 99999. A better method is to use a symbol such as an asterisk (*), or a word, or a sequence of letters (e.g. 'EOD' for 'end of data'). But a 'read' command, which expected a real number, would fail if it were offered a non-numerical character. We need to use a string form of input with a non-numerical marker. Section 13.3.3 illustrates this form of entry.

```
procedure InputAndCountData(var x : DataVector; var n : Units);

(* Data are entered until a marker value shows that entry is *)
(* complete. The data are counted and the total number of    *)
(* data is printed.                                          *)

const marker = -1;

var Value : real;

begin
    n := 0;
```

```
        writeln('Input observations in free format');
        writeln('Input -1 when data entry is complete');
        read(Value);
        if (Value = Marker)
            then writeln('No data entered')
            else begin
                repeat
                    n := n + 1;
                    x[n] := Value;
                    read(Value);
                until (Value = Marker)
            end;
        writeln('Number of data entered = ',n:6)
end; (* of InputAndCountData *)
```

Algorithm 13.1 *InputAndCountData*

13.3   Checking data on entry

13.3.1   Reading checks

The most generally useful check is for the data to be read over carefully by someone who knows what they represent. The basic check we recommend is that at regular intervals, while input proceeds, the data are printed (we 'echo' the data) so that they may be read, checked and corrected. A procedure to do this may be described in pseudo-code as:

repeat
    input block data;
    echo block of data and correct errors
until all data entered;

Algorithm 13.2 (*Input Values*) does this. It includes the procedure *CheckData*. The entry of data is done in a way different from that of Algorithm 13.1. Each entry is prompted by the words 'Item No *i*', with appropriate *i*, and after entering the number, is completed by pressing 'RETURN'. There is no prompt with Algorithm 13.1; the numbers are entered one after the other, separated by a space or a 'RETURN'.

```
procedure InputValues(var x : DataVector; var n : Units);

(* inputs a stated number n of data values which are put in  *)
(* an array x. Values are repeated in blocks of chosen size  *)
(* and may then be amended.                                  *)

const BlockSize = 5;    (* The values of these constants *)
      Field = 10;       (* should be chosen to suit the  *)
      Dp = 2;           (* data being entered.           *)

var i : integer; (* count of values entered *)
    NoInBlock : integer; (* number of values entered in *)
                         (* current block *)
    TotalChecked : integer; (* total number of values    *)
                            (* entered and checked *)
```

```
procedure CheckData( FirstUnit,LastUnit : Units);

var i : integer; (* loop counter *)
    j : integer; (* label of incorrect value *)

begin
    repeat
        writeln('Item No':Field,'Value':Field);
        for i := FirstUnit to LastUnit do
            writeln(i:Field,x[i]:Field:Dp);
        writeln;writeln('Enter item number to be corrected');
        write('If all the values are correct type 0    ');
        readln(j);writeln;
        if j > 0
        then begin
            write('Type correct value for item',j:4,'          ');
            readln(x[j])
        end;
    until j = 0
end; (* of CheckData *)

begin   (* Input of values *)
    i := 0;
    TotalChecked := 0;
    write('State number of values ');readln(n);writeln;
    writeln('Input values. Press "RETURN" after each entry');
    writeln;

    repeat
        NoInBlock := 0;
        writeln('Item No':Field,'Value':Field);writeln;
        while (i < n) and (NoInBlock < BlockSize) do
        begin
            i := i + 1;
            write(i:Field,'        ');readln(x[i]);
            NoInBlock := NoInBlock + 1
        end;
        CheckData(TotalChecked+1,TotalChecked+NoInBlock);
        TotalChecked := TotalChecked + NoInBlock;
    until TotalChecked = n;
    writeln;writeln('Data vector input and check complete');
    writeln
end; (* of InputValues *)
```

Algorithm 13.2 *InputValues*

We give also an algorithm *InputMatrix* (13.3) which is similar to *InputValues* (Algorithm 13.2) but is appropriate for the input of a data matrix instead of a simple list, or vector, of values. There was an example on page 125 with the input of data for the multiple regression program. There were *n* units, and *p* variates measured on each unit. In Algorithm 13.3, the checking is done after the variate values of each unit are entered, using the procedure *CheckRow*. Note that in labelling the elements of the data matrix, the variate index comes before the unit index, e.g. $x[j, i]$ is the value of variate *j* for unit *i*. The procedures may easily be modified for the input of records. (Algorithm 12.1, *InputAnova*, is an example of a simple procedure for the input of records.)

```
procedure InputMatrix(var x : DataMatrix; var n : Units;
                      var nVar : Variates);

(* inputs a stated number n of row vectors each containing   *)
(* nVar elements into a matrix x. Rows are repeated for       *)
(* error checking and amendment where necessary.              *)

const Field = 10; (* The values of these constants should be *)
      Dp = 2;     (* chosen to suit the data being entered.   *)

var i,j : integer; (* loop counters *)


procedure CheckRow;

(* repeats the current row of input and allows rectification *)
(* of errors.                                                *)

var k : integer; (* label *)


begin
    writeln;writeln('Check Unit ',i:3);
    repeat
        writeln;
        for j := 1 to nVar do
            writeln('Variate     ',j:3,x[j,i]:Field:Dp);
        writeln;
        writeln('If all the values are correct type 0');
        writeln('otherwise');
        write('enter number of variate to be corrected   ');
        readln(k);
        if k>0
        then begin
            write('Type correct value for variate ',k:3,'   ');
            readln(x[k,i])
        end;
    until k=0
end; (* of CheckRow *)


begin
    write('Number of Units       ');readln(n);
    write('Number of Variates    ');readln(nVar);
    for i := 1 to n do
    begin
        writeln;writeln;
        writeln('Unit      ',i:3);writeln;
        for j := 1 to nVar do
        begin
            write('Variate ',j:3,'   ');
            readln(x[j,i])
        end; (* of input of one row *)
        writeln;
        CheckRow
    end;
end; (* of InputMatrix *)
```

Algorithm 13.3 *InputMatrix*

13.3.2   Range checks

For many sets of data there is a legitimate range of possible values. Thus examination marks may be known to be in the range 0 to 100. We give Algorithm 13.4, *CheckRange*, which makes this type of check. It sets a boolean variable *RangeError* [*i*] equal to 'true' if an error is found in unit *i*. The procedure may be incorporated into an input routine and used to print an error message at an appropriate time.

```
procedure CheckRange(x : DataVector;
                     FirstUnit,LastUnit : Units;
                     MinValue,MaxValue : real;
                     var RangeError : LogicVector);

(* inspects the block of values of array x from x[FirstUnit] *)
(* to x[LastUnit]. If the i-th value lies outside the range  *)
(* MinValue to MaxValue then RangeError[i] is set to 'true'. *)

var i : integer;

begin
    for i := FirstUnit to LastUnit do
        RangeError[i] := false;
    for i := FirstUnit to LastUnit do
        if (x[i] < MinValue) or (x[i] > MaxValue)
            then RangeError[i] := true;
end; (* of CheckRange *)
```
Algorithm 13.4 *CheckRange*

13.3.3   Errors of input

Errors which cause the most trouble are those produced when the operator, by accident, does something that has not been anticipated. Instead of typing an appropriate digit he may type a letter or press 'RETURN'. To deal with this type of error the response at the keyboard must be input as a string of characters and checked for validity. A foolproof algorithm to do this would be extremely long. We illustrate what is required with Algorithm 13.5, *ReadNumber*: an algorithm for reading an integer number. (In fact, there is a good case for arranging that all input of data be in integers. Including a decimal point in numbers increases the time taken to type them, and increases the chance of error. It is easy to make the computer convert an integer back to a real number by multiplying by the appropriate power of 10.)

```
procedure ReadNumber(var Number:integer);

(* reads input as a string and checks if it is an integer; *)
(* if check fails, requests that an integer be entered.     *)

var Error:boolean;
    i:integer;


procedure  ReadString(var Word:String;var Wordlength:integer);

var Character:char;
    i:integer;
```

```
begin
   Word:=Blank;
   i:=1;
   repeat
      read(Character);
      Word[i]:=Character;
      i:=i+1
   until (Character=' ') or (i>MaxLength);
   if i=1
      then WordLength:=0
      else WordLength:=i-2
end; (* of ReadString *)


begin
   repeat
   ReadString(Word,WordLength);
   if Word=Blank
      then begin
        writeln('Please enter number');Error:=true
      end
      else begin
   Error:=false;
   i:=1;
   Number:=0;
   while Word[i]=' ' do
      i:=i+1;
   while (i<=WordLength) and (Error=false) do
   begin
      if Word[i] in ['0'..'9']
         then Number:=10*Number + (ord(Word[i])-ord('0'))
         else begin
            Error:=true;
            i:=i-1
         end;
      i:=i+1;
     end;
   if Error
      then writeln('Please enter integral number');
      end;
until not Error
end; (* of ReadNumber *)
```

Algorithm 13.5 *ReadNumber*

Algorithm 13.5 includes a subsidiary procedure *ReadString* which reads the input as a string of characters called *Word* and counts the length of the string. If the characters in *Word* represent an integer number, they are converted to the number and put in *Number*, a variable of type integer. Otherwise, the user is asked to re-enter the number.

13.4 Use of files

As we have said earlier, the best way to work with large sets of data is to put the data in a file. We choose to use textfiles only, in our discussion, since (1) they allow general information to be included at the beginning of the file, (2) they may be listed and amended at a terminal. In all the algorithms a complete array of data is transferred to or from the file as a single operation. In some cases less storage would be required in the computer if movement to and from a file were interspersed with calculations, but we recommend keeping file transfers and calculations separate from each other.

13.4.1 Creating a data file

Algorithm 13.6, *FileData*, assumes that a procedure such as *InputMatrix* (Algorithm 13.3) has been used so that a data matrix and the corresponding numbers of units and variates are stored in the computer. *FileData* creates a textfile which contains, in order, (1) the name of the data, which is input as part of the procedure, (2) the number of data units, (3) the number of variates, and (4) the matrix of data. The file is a variable which changes its value when the procedure is run, so the file name must be included as a variable parameter in the parameter list of the procedure. This explains the presence of 'var *DataFile*: text' in the parameter list of *FileData*.

```
procedure FileData(x : DataMatrix; n : Units; nVar : Variates;
                   var DataFile : text);

(* creates a textfile called DataFile containing (1) the name *)
(* of the data, (2) the number of data units, (3) the number  *)
(* of variates and (4) the matrix of data.                    *)

type Name = packed array[1..10] of char;

var DataName : Name;
    i,j : integer;

begin
    rewrite(DataFile);
    write('Enter title of data    ');readln(DataName);
    writeln;
    (* each write statement now writes to DataFile *)
    writeln(DataFile,DataName); (* writes file name *)
    writeln(DataFile,n,nVar); (* writes matrix size *)
    for i:= 1 to n do
    begin
        for j:= 1 to nVar do
            write(DataFile,x[j,i]);   (* writes row i *)
        writeln(DataFile)
    end;
    writeln('DataFile created')
end; (* of FileData *)
```

Algorithm 13.6 *FileData*

Program 13.7, *InputAndStoreData*, combines Algorithms 13.3 and 13.6 into a complete program. Note that the name of the data file (i.e. *DataFile*) must be included, in brackets, after the name of the program; the names 'input' and 'output', also in the brackets, refer to textfiles associated with the main input and output devices which we may take to be, in both cases, the terminal that the reader uses. When we wish to transfer information to a file the name, e.g. *DataFile*, is used as a 'destination' parameter in a write statement. Thus the statement

write (*DataFile, xMean*)

would send the value of the variable *xMean* to *DataFile* instead of to the screen at the terminal. In normal use of 'read' and 'write' the 'destination' files 'input' and 'output', respectively, are understood and need not be included explicitly in the statements. In some implementations of Pascal (or in some operating systems) it is possible to give the name of the data file in response to a question within the program. But this is not standard Pascal and the reader would need to check if this facility is available on his computer. Finally, note one other important requirement: the data file must be opened explicitly by the instruction

rewrite (*DataFile*);

before any attempt is made to write the file.

```
program InputAndStoreData(input,output,DataFile);

(* takes the input of a data matrix from the keyboard, *)
(* allows the user to correct the entries and creates  *)
(* a text file on disc.                                *)

const MaxSampleSize = 100;
      MaxVar = 5;

type Units = 1..MaxSampleSize;
     Variates = 1..MaxVar;
     DataVector = array[Units] of real;
     DataMatrix = array[Variates] of DataVector;

var DataFile : text;
    n : Units;
    nVar : Variates;
    x : DataMatrix;

procedure InputMatrix(var x : DataMatrix; var n : Units;
                      var nVar : Variates);

     (*** Algorithm 13.3 ***)



procedure FileData(x : DataMatrix; n : Units; nVar : Variates;
                   var DataFile : text);

     (*** Algorithm 13.6 ***)



begin
    InputMatrix(x,n,nVar);
    FileData(x,n,nVar,DataFile)
end.
```
Program 13.7 *InputAndStoreData*

We have presented an algorithm for storing a two-dimensional array of data on file. The reader should have no difficulty in modifying the algorithm for other data structures. The most common data structure we have used in this book is that of a set of observations of a single variate, i.e. a vector, or one-dimensional array, of data. This may be regarded as a special case of a data matrix with the number of variates equal to one, but it is probably best to modify the coding and use a variable *x* of type *DataVector*. Another data structure we have used is that of an array of records (Chapter 12). Algorithm 12.1, *InputAnova*, indicates the type of instructions needed when passing records to a file.

13.4.2   Reading from a data file

Reading from a data file which has been created by *FileData* is done using a very similar procedure. The main changes are that the file is made ready for reading by the instruction

reset (*DataFile*);

which replaces the 'rewrite' statement at the beginning of *FileData*, and that 'write' statements are replaced by 'read' statements, (see Algorithm 13.8, *ReadFile*). Files are read in the order in which they are written so the initial information, on name of data and number of units and variates, must always be read even if it is not used.

```
procedure ReadFile(var DataFile : text; var x : DataMatrix;
                   var n : Units; var nVar : Variates);

(* reads the contents of a textfile called DataFile which    *)
(* contains (1) the name of the data, (2) the number of data *)
(* units n, (3) the number of variates nVar and (4) the      *)
(* matrix of data x.                                         *)

type Name = packed array[1..10] of char;

var DataName : Name;
    i,j : integer;

begin
    reset(DataFile);
    readln(DataFile,DataName);  (* reads file name *)
    readln(DataFile,n,nVar);    (* reads matrix size *)
    for i:= 1 to n do
    begin
        for j:= 1 to nVar do
            read(DataFile,x[j,i]);  (* reads row i *)
    end
end; (* of ReadFile *)
```
Algorithm 13.8 *ReadFile*

ReadFile is an algorithm for reading a data matrix from file. It is not difficult to modify it so that it will read other data structures such as a data vector or an array of records.

[*Note*. We give no test data for the algorithms of this chapter, and recommend the reader to use some (or all) of the 'Student data set' (Appendix B).]

13.5 Exercises

1 Modify *Input Values* (Algorithm 13.2) so that the number of data items is not stated but is indicated using a marker.

2 Modify *Input Matrix* (Algorithm 13.3) so that it may be used for the input of records suitable for the analysis of variance (see Section 12.2). Make a further modification so that the treatment name may be entered as a letter, which is converted to an integer for storing in the record.

3 Modify *FileData* (Algorithm 13.6) so that it will transfer records to a data file. Combine this with the input procedure of question 2 to produce a program equivalent to *Input AndStoreData* (Program 13.7).

4 Input of large sets of data into a matrix for subsequent transfer to a file can 'lock up' large amounts of store. Modify procedure *Input Matrix* (Algorithm 13.3) to transfer to the data file each block of data as it is checked. This will have the same effect as *Input AndStoreData* (Program 13.7) but will not hold the whole matrix in memory before transfer to the file.

5 Use the following algorithm to read from a file which has been produced by *FileData* (Algorithm 13.6).

```
procedure InputFile(var DataFile : text; var x : DataMatrix;
                    var n : Units);

type Name = packed array[1..10] of char;

var DataName : Name;
    k : integer; (* counter *)
begin
    reset(DataFile);
    readln(DataFile,DataName);   (* reads file name *)
    n := 0;
    while not eof(DataFile) do
    begin
        n := n + 1;
        k := 0;
        while not eoln(DataFile) do
        begin
            k := k + 1;
            read(DataFile,x[k,n])
        end;
    readln(DataFile)
    end
end;
```

Modify the algorithm so that it checks whether the data conform to a pattern of *n* units each of *nVar* variates, where *n* and *nVar* are given.

6† Construct a file of the 'Student data set' (Appendix B).

7 When entering data for an analysis of variance (particularly if the data set is large) it *may* be considered a waste of space to input treatment and replicate numbers. Provided the data are ordered before entry, only the values need be input, the values for replicates of each treatment being entered in free format, i.e. separated by spaces, and the data for each treatment being terminated by 'RETURN'. If a data file is constructed from data entered in this way, the file may be read back using 'eoln' and 'eof' instructions. At the same time the number of treatments and a vector whose ith element is the number of replicates of treatment i may be determined. Modify the algorithm *InputFile* of question 5 to do this.

Appendix A Ancillary Algorithms

Algorithm A.1 *BetaRatio*

```
function BetaRatio(x,a,b,LnBeta : real) : real;

(* calculates the incomplete beta function ratio with   *)
(* parameters a and b. LnBeta is the logarithm of the    *)
(* complete beta function with parameters a and b.       *)

const error = 1.0E-7;

var c : real; (*   c = a + b   *)
    Factor1,Factor2,Factor3 : real; (* factors multiplying *)
                                    (* terms in series      *)
    i,j : integer; (* counters *)
    Sum : real; (* current sum of series *)
    Temp : real; (* temporary store for exchanges *)
    Term : real; (* term of series *)
    xLow : boolean; (* status of x which determines the end *)
                    (* from which the series is evaluated    *)
    y : real; (* adjusted argument *)

begin
    if (x=0) or (x=1)
    then
        Sum := x
    else begin
        c := a + b;
        if a < c*x
        then begin
            xLow := true;
            y := x;
            x := 1 - x;
            Temp := a;
            a := b;
            b := Temp
        end
        else begin
            xLow := false;
            y := 1 - x;
        end;
        Term := 1;
        j := 0;
        Sum := 1;
        i := trunc(b+c*y)+1;
        Factor1 := x/y;
```

```
        repeat
            j := j + 1;
            i := i - 1;
            if i >= 0
            then begin
                Factor2 := b - j;
                if i = 0 then Factor2 := x;
            end;
            Term := Term*Factor2*Factor1/(a+j);
            Sum := Sum + Term;
        until (abs(Term) <= Sum) and (abs(Term) <= error*Sum);
        Factor3 := exp(a*ln(x) + (b-1)*ln(y) - LnBeta);
        Sum := Sum*Factor3/a;
        if xLow
            then Sum := 1 - Sum;
    end;
    BetaRatio := Sum;
end; (* of BetaRatio *)
```

Algorithm A.2   *InverseBetaRatio*

```
function InverseBetaRatio(Ratio,a,b,LnBeta : real) : real;

(* calculates the inverse of the incomplete beta function  *)
(* ratio with parameters a and b. LnBeta is the logarithm   *)
(* of the complete beta function with parameters a and b.   *)
(* Uses function BetaRatio.                                 *)

const error = 1.0E-7;

var c : real; (* c = a + b *)
    LargeRatio : boolean;
    temp1,temp2,temp3,temp4: real; (* temporary variables *)
    x,x1 : real; (* successive estimates of inverse ratio *)
    y : real; (* adjustment during Newton iteration *)

begin
    if (Ratio = 0) or (Ratio = 1)
    then
        x := Ratio
    else begin
        LargeRatio := false;
        if Ratio > 0.5
        then begin
            LargeRatio := true;
            Ratio := 1 - Ratio;
            temp1 := a;
            b := a;
            a := temp1;
        end;
        c := a + b;
```

```
    (* calculates initial estimate for x *)
    temp1 := sqrt(-ln(Ratio*Ratio));
    temp2 := 1.0 + temp1*(0.99229 + 0.04481*temp1);
    temp2 := temp1 - (2.30753 + 0.27061*temp1)/temp2;
    if (a > 1) and (b > 1)
    then begin
        temp1 := (temp2*temp2 -3.0)/6.0;
        temp3 := 1.0/(a + a - 1.0);
        temp4 := 1.0/(b + b - 1.0);
        x1 := 2.0/(temp3 + temp4);
        x := temp1 + 5.0/6.0 - 2.0/(3.0*x1);
        x := temp2*sqrt(x1 + temp1)/x1 - x*(temp4 - temp3);
        x := a/(a+b*exp(x + x))
    end
    else begin
        temp1 := b + b;
        temp3 := 1.0/(9.0*b);
        temp3 := 1.0 - temp3 + temp2*sqrt(temp3);
        temp3 := temp1*temp3*temp3*temp3;
        if temp3 > 0
        then begin
            temp3 := (4.0*a + temp1 - 2.0)/temp3;
            if temp3 > 1 then x := 1.0-2.0/(1 + temp3)
            else x := exp((ln(Ratio*a) + LnBeta)/a)
        end
        else x := 1.0 - exp((ln((1-Ratio)*b) + LnBeta)/b);
    end;

    (* Newton iteration *)
    repeat
        y := BetaRatio(x,a,b,LnBeta);
        y := (y-Ratio)*exp((1-a)*ln(x)+(1-b)*ln(1-x)+LnBeta);
        temp4 := y;
        x1 := x - y;
        while (x1 <= 0) or (x1 >= 1) do
        begin
            temp4 := temp4/2;
            x1 := x - temp4
        end;
        x := x1;
    until abs(y) < error;
    if LargeRatio then x := 1 - x;
    end;
    InverseBetaRatio := x
end; (* of InverseBetaRatio *)
```

Algorithm A.3 *FindMultiSP*

```
procedure FindMultiSP(x : DataMatrix; n : Units; p : Variates;
                      var Mean : Vector;
                      var SP : MatrixArray);

(* calculates the corrected sums of squares and products for *)
(* p variates from a n*p data matrix x; the results are put  *)
(* in a one-dimensional array SP, with p(p+1)/2 elements,    *)
(* which is the lower triangle, read by rows, of the p*p     *)
(* matrix of sums of squares and products.                   *)

var  i,j,k : integer; (* counters *)
     mj,mk : real; (* mean values of variates *)
     RowMark : integer; (* index of element of SP *)
     Sum : real; (* temporary store for summing values *)
     xj,xk : DataVector; (* set of values of variate j,k *)

begin
    for j := 1 to p do
    begin
        xj := x[j];
        Sum := 0.0;
        for i := 1 to n do
            Sum := Sum + xj[i];
        Mean[j] := Sum/n
    end;
    for j := 1 to p do
    begin
        RowMark := (j*(j - 1)) div 2;
        xj := x[j];
        mj := Mean[j];
        for k := 1 to j do
        begin
            xk := x[k];
            Sum := 0.0;
            mk := Mean[k];
            for i := 1 to n do
                Sum := Sum +(xj[i]-mj)*(xk[i]-mk);
            SP[RowMark+k] := Sum
        end (* of loop indexed by k *)
    end (* of loop indexed by j *)
end; (* of FindMultiSP *)
```

Algorithm A.4 *LinearSolver*

```
procedure LinearSolver(A : MatrixArray ; h : Vector;
                       NoOfEqns : integer;
                       var x :Vector;
                       var AInverse : MatrixArray;
                       var Nullity : integer);

(* solves the system of linear equtions A*x=h where A is a     *)
(* symmetric positive definite matrix, using the procedures    *)
(* Factorise and Solve. Factorise finds a lower triangular     *)
(* matrix F such that F*F' = A. Solve performs forward and     *)
(* backward substitution to find the solution vector x.        *)
(* In addition the inverse of A is found by the procedure      *)
(* Invert.  A,F and AInverse are stored as lower-triangular    *)
(* matrices in row order.                                      *)

const accuracy = 1.0E-6;

var ArraySize : integer; (* number of elements in A etc *)
    Element : real; (* an element of an array *)
    F : MatrixArray; (*  the factor of A such that F*F' = A *)
    i,j,k : integer; (* loop counters *)
    p,q : integer; (* labels *)

procedure Factorise;

begin
    for i := 1 to ArraySize do
        F[i] := A[i];
    Nullity := 0;
    for j := 1 to NoOfEqns do
    begin
        p := (j*(j+1)) div 2;
        if j > 1
        then begin
            for i := j to NoOfEqns do
            begin
                q := (i*(i-1)) div 2 + j;
                Element := F[q];
                for k := 1 to j-1 do
                    Element := Element - F[q-k]*F[p-k];
                F[q] := Element
            end
        end;
        if F[p] < accuracy
        then begin
            F[p] := 0;
            Nullity := Nullity + 1
        end;
        Element := sqrt(F[p]);
        for i := j to NoOfEqns do
        begin
            q := (i*(i-1)) div 2 +j;
            if Element = 0
                then F[q] := 0
                else F[q] := F[q]/Element
        end
    end;
end; (*  of Factorise   *)
```

```
procedure Solve;

begin
    for i := 1 to NoOfEqns do
        x[i] := h[i];
    if F[1] = 0
        then x[1] := 0
        else x[1] := x[1]/F[1];
    if NoOfEqns > 1
    then begin
        k := 1;
        for i := 2 to NoOfEqns do
        begin
            Element := x[i];
            for j := 1 to i-1 do
            begin
                k := k + 1;
                Element := Element - F[k]*x[j];
            end;
            k := k + 1;
            if F[k] = 0
                then x[i] := 0
                else x[i] := Element/F[k]
        end
    end;
    if F[ArraySize] = 0
        then x[NoOfEqns] := 0
        else x[NoOfEqns] := x[NoOfEqns]/F[ArraySize];
    if NoOfEqns >1
    then begin
        for i := NoOfEqns downto 2 do
        begin
            Element := x[i];
            k := (i*(i-1)) div 2;
            for j := 1 to i-1 do
                x[j] := x[j] - Element*F[k+j];
            if F[k] = 0
                then x[i-1] := 0
                else x[i-1] := x[i-1]/F[k]
        end
    end
end; (*    of Solve    *)


procedure Invert;

var iCol,iRow,jCol,nDiag,nRow : integer; (* labels *)
    T : Vector; (* temporary array *)

begin
    for i := 1 to ArraySize do
        AInverse[i] := F[i];
    nRow := NoOfEqns;
    iRow := nRow;
    nDiag := ArraySize;
```

```
    repeat
        if AInverse[nDiag] = 0
        then begin
            p := nDiag;
            for j := iRow to nRow do
            begin
                AInverse[i] := 0;
                p := p + j;
            end
        end
        else begin
            p := nDiag;
            for i := iRow to nRow do
            begin
                T[i] := AInverse[p];
                p := p + i
            end;
            iCol := nRow + 1;
            jCol := ArraySize + 1;
            q := ArraySize + iCol;
            repeat
                q := q - iCol;
                iCol := iCol - 1;
                jCol := jCol - 1;
                p := jCol;
                Element := 0.0;
                if iCol = iRow
                    then Element := 1.0/T[iRow];
                k := nRow;
                while k > iRow do
                begin
                    Element := Element - T[k]*AInverse[p];
                    k := k - 1;
                    p := p - 1;
                    if p > q
                        then p := p - k + 1;
                end;
                AInverse[p] := Element/T[iRow];
            until iCol = iRow
        end;
        nDiag := nDiag - iRow;
        iRow := iRow - 1
    until iRow = 0
end; (*  of Invert   *)



begin      (*   LinearSolver   *)
    ArraySize := (NoOfEqns*(NoOfEqns+1)) div 2;
    Factorise;
    Solve;
    Invert
end; (*   of Linearsolver   *)
```

Algorithm A.5   *OrderRecords*

```
procedure OrderRecords(var DataSet:DataRecords; var n:Units);

(* orders the records entered by InputAnova first by          *)
(* treatment and then by replicate within each treatment      *)
(* group.  Each record in DataSet is assigned a 'place'.       *)
(* This placing is ordered using an extension of ShellSort.   *)
var Gap : integer; (* gap between placevalues being compared *)
    i,j : integer; (* labels and counters *)
    Item : LineOfData; (* store for exchanging records *)
    Nextj : integer; (* j + Gap *)
    Place : DataList; (* vector of placevalues *)
    PlaceValue : integer; (* store to exchange place values *)

begin
    for i := 1 to n do
    with DataSet[i] do
       Place[i] := TrNo*n + RepNo;
    Gap := n;
    repeat
        Gap := Gap div 2;
        if Gap > 0
        then begin
            for i := 1 to n-Gap do
            begin
                j := i;
                while j >= 1 do
                begin
                    Nextj := j + Gap;
                    if Place[j] > Place[Nextj]
                    then begin
                        PlaceValue := Place[j];
                        Place[j] := Place[Nextj];
                        Place[Nextj] := PlaceValue;
                        Item := DataSet[j];
                        DataSet[j] := DataSet[Nextj];
                        DataSet[Nextj] := Item
                    end (* of exchange *)
                    else j := 0;
                    j := j - Gap
                end (* of while loop *)
            end (* of comparison for given i *)
        end; (* of pass with given gap size *)
    until Gap = 0
end; (* of OrderRecords *)
```

Appendix B Student data set

We give here a large data set, collected from 100 university students. The first column is simply the student reference number; the next four columns are self-explanatory (but note that 'years' is quoted as a decimal, not as years and months). The variable 'left-handedness' was recorded on an integer scale from 0 to 20, a high score indicating a very left-handed person based on answers to a questionnaire which listed various activities (e.g. writing, throwing, striking a match) that can be done with either hand. 'Reaction time' was the length of time that elapsed between a light being flashed and the person pressing a button. 'Card sorting time' was the time taken to sort a thoroughly shuffled standard pack of playing cards into the four suits.

References in the text have already suggested uses for parts of these data, and further suggestions follow the data (page 164).

B.1 Student data set

S   Sex, M or F
H   Height in centimetres
W   Weight in kilograms
A   Age in years
L   Lefthandedness score, 0–20
R   Reaction time in milliseconds
C   Card sorting time in seconds

Number	S	H	W	A	L	R	C
1	M	176	72	22.6	0	212	33
2	M	172	58	18.5	20	163	56
3	M	177	60	19.1	3	220	50
4	M	174	68	19.6	0	169	66
5	M	167	59	18.9	8	140	62
6	M	191	72	19.9	7	156	63
7	M	190	85	19.4	3	178	95
8	M	176	73	19.0	6	136	52
9	F	165	68	18.2	4	205	28
10	F	167	63	18.0	3	194	48
11	F	172	60	18.5	6	186	51
12	F	169	58	19.6	3	172	63
13	M	173	57	18.9	1	149	71
14	M	180	76	19.2	2	171	57
15	F	154	49	19.9	6	198	59
16	M	178	69	20.1	5	182	72

Number	S	H	W	A	L	R	C
17	M	170	74	19.8	8	257	60
18	M	173	63	19.0	3	155	83
19	M	168	50	21.1	1	139	62
20	M	173	64	20.9	3	137	73
21	M	178	71	19.0	2	162	34
22	F	170	63	19.2	17	176	64
23	F	175	58	18.9	5	175	62
24	F	156	57	20.6	7	207	98
25	F	170	55	18.3	4	169	60
26	M	182	66	18.4	3	226	55
27	M	175	67	18.9	3	171	50
28	F	173	57	19.4	4	188	59
29	M	175	65	20.0	11	177	58
30	M	173	61	19.5	5	148	55
31	M	178	68	19.7	18	131	64
32	M	174	64	19.3	4	263	65
33	M	176	66	29.1	6	159	59
34	F	177	65	22.9	6	156	49
35	F	161	47	20.1	15	193	60
36	M	167	64	18.5	4	140	62
37	F	168	59	33.9	3	162	55
38	M	185	76	34.2	18	167	58
39	F	164	67	24.1	7	254	72
40	F	160	55	29.4	1	277	51
41	M	174	55	18.5	5	172	51
42	F	154	52	18.5	6	231	55
43	M	170	60	20.4	5	186	58
44	M	176	68	19.1	3	174	56
45	M	160	53	22.3	14	176	52
46	M	174	71	28.3	18	153	66
47	F	148	33	19.9	4	172	62
48	F	161	53	25.2	4	205	55
49	M	182	72	28.0	4	210	58
50	M	184	80	19.2	8	201	60
51	F	158	53	19.0	6	187	55
52	F	170	54	18.5	1	162	62
53	F	157	50	19.8	0	145	61
54	F	173	52	18.8	10	95	53
55	F	162	70	18.2	8	207	61
56	F	163	53	19.9	7	159	57
57	M	183	73	17.9	0	220	54
58	M	179	68	19.0	0	220	62
59	M	164	63	23.6	6	160	51
60	M	177	72	18.8	15	260	60
61	M	179	66	23.1	3	160	64
62	F	172	64	18.4	6	240	55
63	F	174	60	18.7	3	170	52
64	F	173	68	18.9	4	175	56
65	F	164	57	18.9	3	245	56
66	M	184	74	20.1	0	275	63
67	M	181	64	20.0	7	200	54
68	F	164	52	19.0	4	176	67

Number	S	H	W	A	L	R	C
69	M	177	67	27.0	0	210	54
70	M	177	69	22.9	8	193	53
71	M	178	73	18.2	5	162	59
72	M	178	71	18.9	9	280	62
73	M	181	60	19.1	9	220	63
74	F	157	49	19.9	19	160	55
75	M	182	70	18.9	8	180	61
76	M	183	63	18.9	0	140	60
77	M	184	92	24.9	4	155	78
78	F	170	57	19.0	15	350	50
79	F	168	54	18.5	2	375	57
80	F	159	53	18.4	3	340	65
81	F	157	49	19.0	5	205	69
82	F	160	50	18.4	7	175	60
83	M	176	61	18.0	3	191	62
84	F	162	52	18.9	6	168	67
85	F	161	47	21.2	5	220	54
86	F	168	53	19.2	3	210	68
87	M	170	66	18.6	7	220	60
88	M	174	67	20.4	7	200	53
89	F	155	49	19.7	1	170	62
90	M	160	52	19.5	8	220	72
91	M	172	64	19.5	9	230	75
92	F	166	58	18.4	8	235	42
93	F	161	64	18.8	5	250	33
94	F	160	51	18.6	2	184	47
95	M	191	96	18.8	6	190	67
96	M	174	52	22.0	2	140	52
97	F	168	52	18.9	4	155	74
98	F	176	70	19.1	1	262	53
99	F	162	49	18.9	4	238	63
100	F	161	49	18.8	4	171	61

B.2   Exercises using student data set

Chapter 3   Sorting and ranking
1 Sort on H or W or A or L or R or C (any sorting method).
2 List the student numbers in order of age.
3 Exercises 3.9, question 10: Run using H or W or R separately for each sex.
4 Exercises 3.9, question 13: Run using any column of the data but separating the sexes.

Chapter 4   Inspection and summary of data using tables
1 Exercises 4.9, question 3: Run using (i) L in classes of unit width; (ii) H in classes of width five.
2 Exercises 4.9, question 5: Run using any column of the data but separating by sex.
3 Run Program 4.6 (*ContingencyTable*) with the sex and lefthandedness columns. Convert the lefthandedness score into a hand code: L corresponds to a score greater than 10 and R to a score less than or equal to 10.
4 Exercises 4.9, question 6: Run with $(x, y) = (H, W)$, separating the sexes.
[*Note*: be careful in choice of class widths.]
5 Exercises 4.9, question 6: Run with $(x, y) = (S, L)$, with L grouped into (0, 5), (6, 10), (11, 15), (16, 20).

Chapter 5   Inspection and summary of data using graphical methods
1 Produce histograms for each of the columns.
2 Exercises 5.10, question 2: Run using each column of the data.
[*Note*: Ages should be converted to 3-digit integers.]
3 Exercises 5.10, question 4: Run using each column of the data but separating the sexes, and printing the plot for one sex directly underneath the plot for the other sex.

Chapter 6   Computation of variance and correlation coefficients
1 Calculate and print out the mean and variance of each column of the data, separating the sexes.
2 Calculate and print out the correlation coefficient of the following pairs of variables: (H, W), (W, A), (L, C), (R, C). (Separate the sexes for the first two pairs.)
3 Exercises 6.10, question 9: Run with the pairs of variables suggested in (2) above.

Chapter 10   Significance tests and confidence intervals
1 Carry out a chi-squared test on the contingency table produced above in Exercise 3 for Chapter 4.
2 Exercises 10.7, question 10: Test whether (i) heights, (ii) weights are normally distributed for each sex.
3 Compare (i) reaction times R and (ii) card sorting times C for the two sexes using:
(a) an unpaired *t*-test (Section 10.4.1 and page 120, question 13);
(b) *FindRankSumStatistic* (Algorithm 10.4).

Bibliography

ABRAMOWITZ, M, and STEGUN, I A, *Handbook of Mathematical Functions*. Dover, New York (1972)

ASHBY, T, A modification to Paulson's approximation to the variance ratio distribution. *The Computer Journal*, **11**, 209–10 (1968)

CHAMBERS, J M, Computers in statistics. *American Statistician*, **34**, 238–43 (1980)

CHAN, T F, GOLUB, G H, and LEVEQUE, R J, Algorithms for computing the sample variance: analysis and recommendations. *American Statistician*, **37**, 242–7 (1983)

CHATTERJEE, S, and PRICE, B, *Regression Analysis by Example*. Wiley, New York (1977)

CLARKE, G M, *Statistics and Experimental Design*. 2nd Edition, Edward Arnold, London (1980)

CLARKE, G M, and COOKE, D, *A Basic Course in Statistics*. 2nd Edition, Edward Arnold, London (1983)

CRAN, G W, MARTIN, K J, and THOMAS, G E, A remark on algorithms AS 63 and AS 64. *Applied Statistics*, **26**, 111–14 (1977)

DRAPER, N R, and SMITH, H, *Applied Regression Analysis*. 2nd Edition, Wiley, New York (1981)

FISHMAN, G S, *Priciples of Discrete Event Simulation*. Wiley, New York (1978)

GEARY, R C, The frequency distribution of the quotient of two normal variables. *Journal of the Royal Statistical Society*, **93**, 442 (1930)

HOGG, R V, and CRAIG, A T, *Introduction to Mathematical Statistics*. 3rd Edition, Macmillan, New York; Collier-Macmillan, London (1970)

JOHNSON, L W, and RIESS, R D, *Numerical Analysis*. 2nd Edition, Addison-Wesley, Reading, Mass. (1982)

KNUTH, D E, *The Art of Computer Programming*. Vol. 2, Seminumerical Algorithms. 2nd Edition, Addison-Wesley, Reading, Mass. (1981)

KNUTH, D E, *The Art of Computer Programming*. Vol. 3, Sorting and Searching. Addison-Wesley, Reading, Mass. (1973)

LAU, C L, Algorithm AS 147. A simple series for the incomplete gamma integral. *Applied Statistics*, **29**, 113–14 (1980)

LORIN, H, *Sorting and Sort Systems*. Addison-Wesley, Reading, Mass. (1975)

MAJUMDER, K L, and BHATTACHARJEE, G P, Algorithm AS 63. The incomplete beta integral. *Applied Statistics*, **22**, 409–11 (1973a)

MAJUMDER, K L, and BHATTACHARJEE, G P, Algorithm AS 64. Inverse of the incomplete beta function ratio. *Applied Statistics*, **22**, 411–14 (1973b)

MORGAN, B J T, *Elements of Simulation*, Chapman and Hall, London (1984)

PAULSON, E, An approximate normalization of the analysis of variance distribution. *Annals of Mathematical Statistics*, **13**, 233–5 (1942)

RIPLEY, B D, Computer generation of random variables: a tutorial. *International Statistical Review*, **51**, 301–19 (1983)

SCHNEIDER, G M, and BRUELL, S C, *Advanced Programming and Problem Solving with Pascal*. Wiley, New York (1981)

SNEDECOR, G W, and COCHRAN, W G, *Statistical Methods*. 7th Edition, Iowa State University Press, Ames, Iowa (1980)

TOCHER, K D, *The Art of Simulation*. English Universities Press, London (1963)

TUKEY, J W, *Exploratory Data Analysis*. Addison-Wesley, Reading, Mass. (1977)

WALLACE, D L, Bounds on normal approximations to Student's and the chi-squared distributions. *Annals of Mathematical Statistics*, **30**, 1121–30 (1959)

WILKINSON, J H, *Rounding Errors in Algebraic Processes*. HMSO, London (1963)

WILSON, E B, and HILFERTY, M M, The distribution of chi-square. *Proc. National Academy of Sciences*, **17**, 684–8, Washington (1931)

WILSON, I R, and ADDYMAN, A M, *A Practical Introduction to Pascal — with BS* 6192. 2nd Edition, Macmillan, London (1982)

WIRTH, N, *Algorithms + Data Structures = Programs*. Prentice-Hall, Englewood Cliffs, NJ (1976)

YAKOWITZ, S J, *Computational Probability and Simulation*. Addison-Wesley, Reading, Mass. (1977)

Index